Strategic Thinking Illustrated

This book is about the behaviour of systems. Systems are important, for we interact with them all the time, and many of the actions we take are influenced by a system – for example, the system of performance measures in an organisation influences, often very strongly, how individuals within that organisation behave. Furthermore, sometimes we are involved in the design of systems, as is any manager contributing to the definition of what those performance measures might be. That manager will want to ensure that all the proposed performance measures will drive the 'right' behaviours rather than (inadvertently) encouraging dysfunctional 'game playing', and so anticipating how the performance measurement system will work in practice is a vital part of a wise design process.

Some of the systems with which we interact are local, such as your organisation's performance measurement system. Some systems, however, are distant, but nonetheless very real, such as the healthcare system, the education system, the legal system and the climate system. Systems therefore exist on all scales, from the local to the global. And all systems are complex, some hugely so. That's why understanding how systems behave can be very helpful.

Systems are complex for two main reasons. First, the manner in which they behave over time can be very hard to anticipate – and anticipating the future sensibly is of course a key objective of management. Second, the 'entities' within a system can be connected together in very complex ways, so that an intervention 'here' can result in an effect 'there', perhaps a long time afterwards. Sometimes this can be surprising, and so we talk of 'unintended consequences' – but this is of course a euphemism for 'because I didn't understand how this system behaves, I had not anticipated that'.

Systems thinking, the subject matter of this book, is the disciplined study of systems, and causal loop diagrams – the 'pictures' of this 'picture book' – are a very insightful way to represent the connectedness of the entities from which any system is composed, so taming that system's complexity.

Strategic Thinking Illustrated

Strategy Made Visual Using Systems Thinking

Dennis Sherwood

The Silver Bullet Machine Manufacturing Company Limited

A PRODUCTIVITY PRESS BOOK

First published 2023
by Routledge
605 Third Avenue, New York, NY 10158

and by Routledge
4 Park Square, Milton Park, Abingdon, Oxon, OX14 4RN

Routledge is an imprint of the Taylor & Francis Group, an informa business

ISBN: 978-1-032-30234-8 (hbk)
ISBN: 978-1-032-30233-1 (pbk)
ISBN: 978-1-003-30405-0 (ebk)

DOI: 10.4324/9781003304050

Typeset in Garamond
by codeMantra

Contents

PART 2 APPLICATIONS

PART 3 OVER TO YOU!

Foreword

This is a book about thinking strategically using systems thinking. Put simply, 'systems thinking' means seeing the world as a complex, interconnected place. When Dennis Sherwood and I first got interested in systems thinking that was still a novel insight. It is noticeable – and welcome – that today the insight is mainstream. I remember doing some work on the operation of the Accident and Emergency department in a London hospital in the 1990s and realising that charging in with 'My goodness, have you people realised that you are living in a complex, interconnected system?!' would not really get you very far. They knew that already. People in health care have that insight because they are clever and because they spend their lives dealing with one of the most complex, interconnected systems around: the human body. But that insight is not enough. Although it has become mainstream – almost to the point of management cliché – that is not the end of the story.

The issue today is what to do with that insight. What people need is a vehicle that helps them do something with that idea, a tool that helps them move forward. Tools are fine things and we should never be ashamed of reaching for them. The human mind can imagine that it would be useful to fasten two pieces of wood together with a nail. We are also bright enough to know that trying to bring this about with just your fist will just produce a bloody mess. So we invented the hammer. It is a good tool to reach for. So is systems thinking.

The term 'systems thinking' means different things to different people. It is sometimes used for the range of approaches that make up the field of Systems Science. Peter Checkland's 'Soft Systems Methodology' fits with this, as do other approaches that aim to take a holistic view of the world. However, the type of systems thinking discussed here derives from System Dynamics, the creation of Prof. Jay W. Forrester. This concerns itself with causal connections, accumulations and feedback loops and how these can be used to explain how things evolve over time.

Research tells us that, unassisted, people are not that good at thinking about long chains of consequences, of pondering feedback effects, of considering what will happen if they take a certain course of action, enact a given policy. Strategic thinking becomes counter-intuitive, challenging, overwhelming. That is where systems thinking comes in. That is where this book comes in. Systems thinking is the vehicle that people need, the tool that converts the insight of complexity and inter-connectedness into an approach that is practical and helpful and positive.

If you have a suspicion that it is all getting a bit complex, that holding inside your head all the different connections that you think might be there in the world is making your brain feel fluffy, then this book is for you. Simply getting a blank piece of paper and mapping out your ideas gives you a picture of what you are thinking. In cognitive terms it acts as a form of 'virtual memory': you can work on one part of the map, then shift your attention to another part, sure in the knowledge that everything is still there, recorded and ready to be gone back to. You can think about long chains of effects. You might even discover unintended consequences of an action or policy. (Because in systems thinking there are no 'side effects'; rather there are effects that you had not anticipated.) The map can help with that. It can also give you some help with thinking about how a system will evolve over time, what trajectories the value of the variables will have as time flows forward. To do that you need to find the reinforcing and balancing feedback effects and to puzzle out how they interact and drive behaviour. This is the route to strategic thinking.

At some point a qualitative map may not do all that you need. A model with numbers, a model that allows you to press <Run>, a model that simulates the consequences of your assumptions might be useful. Systems thinking includes that too. Simulation modelling requires more work, more effort. But as our colleague Kim Warren likes to say, what is a senior decision-maker's job if not to work hard, to put serious effort into choosing a course of action? Something that will rigorously and unambiguously deduce the consequences over time of your assumptions is a good thing to have to hand. Indeed, I would add 'What if?' to Kipling's 'six honest serving men' to form a septuplet.

There is something else that systems thinking is very, very good for. In Brandi Carlile's elegiac song *That Year*, the narrator describes the decade of anger she has felt since her teenage years when one of her friends died. Now, she tells the listener, her view has shifted. With time and reflection she is able to see him as part of her life, to think of him again as a friend. Although she can explain the wasted years (in terms of immaturity, anger, lack of empathy), she poignantly concludes, in the song's summarising final line, 'I was wrong'. That line always brings me up short, never fails to move me. It expresses a powerful, rich, precious insight. Admitting to being wrong is hard. A training in mathematics exposes you to situations where you are just plain wrong and have to deal with it. Perhaps that helps a little – but only a little. Imre Lakatos describes in philosophical terms the twists and turns we take to guard our core assumptions, to avoid admitting that we are wrong. Even mathematicians. In fact, experiments in behavioural science confirm how we all resist such realisations. But admitting that you are wrong creates a moment of potential. You have the opportunity to change your mind.

There is something particularly compelling about seeing a simulation model produce behaviour over time quite different from what was expected. Having that expectation is key. It is enjoyable to be able to press <Run> and see the time plots flow out. However, to get real benefit it is important to first stop and think, 'What do I believe will happen?'

Remember that expectation, that intuition. Better yet, write it down, or tell a colleague. Then press the button. Because only then can a model truly surprise, actually challenge your intuition. Only then can a model force you to realise that you are wrong. That is the beginning of learning.

Perhaps this makes too much of a distinction. Moving to simulation is not a step-change. A qualitative map or model can still reveal something new, cause a shift in thinking. With a pencil, a rubber, a piece of paper and your brain, you can surprise yourself with your ability to see new things. I still recall my earliest experiences of using causal loop diagramming, how the process of diagramming helped me get new insights into things I thought I had already understood quite well. As I filled the blank pages with variables and connections, curiosity and puzzlement turned into interest and understanding as surprising insights changed my 'mental model' of the world and led me to a more ordered and harmonious way of thinking about complexity and interconnectedness, about 'what ifs' and behaviour over time. I get the same feelings today.

All of which might make systems thinking sound like a rather solitary discipline. It certainly is a discipline, with rules worth learning. However, it is more powerful, more enjoyable than that. Because a map or a simulation model can also be made by a group. When you do this, if you get it right, a number of new and rather wonderful things can happen. The model – I will call both that from hereon – becomes a creative tool. It can help people in a team express their ideas clearly. It can help those around criticise those ideas. It can support debate. It can provoke people to collect empirical data and see how that compares with the thinking of those in the group. Although he was not talking about models as such, John Stuart Mill assured us that activities like this are more likely to lead to bad ideas being dropped or good ideas getting better and justly attracting more confidence. If a group is open in its discussions and is committed to getting a good outcome then this is to be welcomed. Knowledge is elicited, represented and clarified. The model becomes a shared description of reality, a device that captures what people think and why they believe that a certain policy is the right one.

People from a wide range of backgrounds can be drawn into such an approach. I have worked with refinery engineers, marketers, public health officials and youth social workers. Each presents challenges. Those not familiar with formal modelling can be suspicious of the idea that a model can 'capture' their thinking. It is important to assure those people that a model can never contain all of their knowledge, that it does not replace judgement. Rather, it is a tool that supports, that contributes to, their judgement. In contrast, those who are familiar with modelling sometimes want to put too much into a model, or to use the natural sciences as their gold standard. On the first point, 'modelling a system' is a guarantee of deadlock because it has no end. There is no criterion for leaving things out – so it is easy for everything to get thrown in and for the work to get bogged down. Modelling is the art of leaving things out. The solution is to model a problem, not a system. It is still hard to decide what to include but there is some yardstick available for deciding if an effect really needs to be included or whether – for now, because these judgements can always be revisited – we can concentrate on other things. On the second point, whilst no model should be inconsistent with empirical data and evidence, we are not doing physics in a lab. Those who have experienced decision-making in organisations know that a little structure, a little disciplined thinking, can add immeasurably to the quality of a group process. Making things better, not trying to make them perfect, is the aim.

Done right, systems thinking in a group creates things that are much more significant than a model. It creates learning. It creates insight. It creates understanding. It creates commitment to a course of action. The ideal outcome of any systems thinking work is not a model. Rather, it is a group of people who rise from the table with a shared understanding of what they are going to do and what its consequences will be. They leave the room thinking, 'We are going to make this happen. I'm doing this part, Chris is doing that part and it all make sense and it really should work. You want me to explain why? Certainly, sit down and give me a minute or two. But then I want to make this happen…'.

It is a reassuring and ultimately empowering way of working. I am not quite sure when I first became a systems thinker. Sitting in a room in Shell Centre with Barry Richmond's superb manual for the System Dynamics software STELLA is an obvious point. However, as I look back, I think that a doctorate in mathematical biology was an earlier foray. I imagine that Dennis, with his background in physics, biology and biophysics, faces the same uncertainty. I have concluded that systems thinking appeals to something deeper, something perhaps separate from any specific discipline that one might have been trained in. It is almost an intuition of connectivity, an embracing of nuance, an urge. This realisation crystallised for me some years ago when I was teaching at the London School of Economics and Political Science. Some weeks into my course one of my students enthusiastically said to me, 'This is the way that I have always wanted to think. Except I did not know that THIS was the way I had always wanted to think'. There it is – the urge.

If all of this seems appealing and attractive then you are holding in your hand a way of learning more. For two decades now I have had the pleasure of arranging for Dennis to give guest lectures to my students. Masters students at LSE and at Henley Business School have seen him combine rigour, creativity, knowledge and wit, all to apply systems thinking to a range of phenomena. This book brings his approach to a wider audience.

Dennis has chosen a characteristically direct and compelling approach. His 'picture book' comes in easy, bite-sized pieces. The use of colour and the landscape orientation and size of the pages help make it an easy and fun read. But as you go through it you will begin to see the wealth of ideas that are adroitly but clearly laid out, the unifying concepts that are the warp and weft of the whole project. There is rigour and discipline here but offered in a digestible and enjoyable way. There is mapping but also some simulation. And there is the range of applications. In my time I have worked on oil and gas production, Norovirus transmission and child protection. Whisper it softly but one of the deepest joys of systems thinking is its generality, the way in which it can be applied to so many phenomena, giving us the chance to get involved in a wide variety of projects. It is so rewarding. It is so much fun. Actually, we should shout it very loud. This important idea is conveyed here as Dennis draws on his long years of experience to engage with a wide range of business problems, the climate crisis, performance measure for teachers and more. As a closing grace note he offers a piece of systems thinking about the benefits of using systems thinking!

This book will teach you how to do system thinking. When you have read it you will be able to look up and face seemingly tangled problems with confidence. The situation is well described by Stephen Sondheim; his musical *Sunday in the Park with George* contains a celebration of disciplined creativity:

> White, a blank page or canvas
> The challenge, bring order to the whole
> Through Design
> Composition
> Tension
> Balance
> Light
> And Harmony.

Dennis shows how this can be accomplished using systems thinking. Systems thinking is powerful. Systems thinking is persuasive and effective. Systems thinking is also fun. Enough from me. I urge you to accept Dennis's guidance and try it yourself.

Prof David C Lane BSc MSc DPhil FORS

January 2022

David Lane is currently Professor of Business Informatics at Henley Business School in England. He has a BSc degree in Mathematics from the University of Bristol and an MSc and DPhil from the University of Oxford. He specialises in the theory and practice of System Dynamics modelling. He is a recipient of the System Dynamics Society's Jay Wright Forrester Award and its Applications Award, and of the Operational Research Society's President's Medal. He is a Fellow of the Operational Research Society. He is a working class academic.

Preface

Real systems – from the budgeting system in your own organisation to the system underlying your country's provision of health and social care – are complex. And because they are complex, they are difficult to understand, to design, to manage. The challenge, then, is to tame that complexity, for only then can understanding be clarified and deepened; only then can systems be designed that will operate more effectively; only then can managers manage those systems more wisely.

I've been involved in all aspects of systems over the last 40 years, and I have no doubt that by far the most insightful way to tame the complexity of real systems is to use the methodology known as 'system thinking', and to compile 'maps' – known as 'causal loop diagrams' – showing how systems work. The purpose of this book is therefore to introduce and describe systems thinking, and to build confidence in using, reading and compiling casual loop diagrams. Accordingly, this book is intended to be of interest, and value, not only to students but also – and perhaps more importantly – to managers and policy makers who are in the middle of complex systems and trying to manage them as best they can, as well as those who are re-designing existing systems or designing new ones.

And, as a glance will tell, this book is 'different'. Rather than being page after page of dense text, this book is largely a sequence of 'pictures', these being causal loop diagrams which progressively tell increasingly more complex stories. Furthermore, each page is complete in its own right, in that any one page can be read on its own, and is fully intelligible. And for the more complex diagrams, when the print version is laid flat, the diagram is on the left and the corresponding narrative on the right, so forming a two-page spread. Certainly, when read in sequence, successive pages tell a more joined-up story, but this format has been designed for the reality that most of us are very busy, and don't have the time to read lots of text. So you can dip in and out, reading one page or several, whatever works for you.

The book starts with a very simple example, and progressively builds to an analysis of the most complex system there is – the system that underpins global warming and climate change. And at the end, there are some guidelines as to how you can compile your own causal loop diagrams to describe your own systems, and how those diagrams can be used. For systems thinking is fundamentally a practical skill, a skill that really helps tame complexity.

Many of the examples in this book are based on work I have done for clients in all sectors and of all scales over the last 30 years or so. My thanks, of course, to all those clients, especially to the Directors of GPDF Limited as regards the content of Chapter 15, as well as to the late Dr James Lovelock for his inspiration for Chapter 17 and for granting permission to reproduce his materials. I have also had the great benefit of working with, and learning from, many highly talented people, so it is my pleasure to acknowledge, with gratitude, in particular Bernard Minsky, Jim Mather, Professor John Morecroft, John Speed, John Taylor, Professor Markus Schwaninger, Michael Ballé, Professor Michael Kennedy, Kerry Turner, Tom Ilube CBE and Professor David Lane, to whom I give extra-special thanks for contributing the Foreword.

And my thanks too to you, the reader. I trust you will find this book helpful, and will enjoy reading it. And if you do, and if you find the very 'different' format engaging, please join me in thanking Michael Sinocchi and the publishing team at Taylor & Francis and Routledge, and also the production team at codeMantra, who made it all happen.

A Note on Causal Loop Diagrams

This book is full of 'pictures' – causal loop diagrams that describe how a given system behaves. A causal loop diagram, of course, is not the actual system itself; rather, it is a representation of that system, a representation that – I trust – is clear, meaningful and informative. This concept is very familiar: a map is not the actual landscape, but a meaningful representation. And in compiling a map, the geographer does two things.

First, symbolism. Maps use a variety of symbols, often following well-established conventions, to represent different features – for example, a blue line is likely to represent a river, a red line a major road. Anyone unfamiliar with these symbols is likely, at first sight, to find a map confusing and hard to understand. But very quickly, the meaning of each symbol becomes clear, and the consistency of their use on different maps enables them to be learnt so that new maps can be 'read' with confidence.

Second, selection. The map does not show every feature of the landscape, for the map-maker has selected those aspects of the totality that are considered to be especially relevant – the roads and towns for maps to aid driving; places of interest on tourist maps; contours and footpaths on maps for walkers.

Causal loop diagrams also make use of standard symbols. At first sight, a causal loop diagram – such as that shown on page 250 – appears to be a tangled mess of curly arrows, blobs, explosions and words, all in garish colours. And a glance at another – see, for example, page 288 – shows a similar, but different, apparent muddle. Those curly arrows, blobs, explosions and words, however, are the equivalent of the symbols on a road map – symbols that are readily learnt, as will be explained in the first few chapters of this book.

Causal loop diagrams are selective too. The world in which we all live is hugely complex, and the purpose of any causal loop diagram is to capture the essence of a particular aspect of that world, and to tame that particular complexity. In so doing, the compiler of that causal loop diagram inevitably selects some features and ignores others. What to include and what to leave out are matters of personal judgement, judgement strongly influenced by the 'mental model' of the compiler, as discussed on page 9. And different compilers have different mental models, and so may compile different diagrams of the same reality.

I mention that now, for all the diagrams in this book have been compiled by me; they are my mental models. As you read them – especially those in the latter chapters that describe more complex realities – you may disagree: perhaps you might think that something I have included is superfluous, and that something you think important is missing. In which case, that's fine – please take my diagram and amend it as you wish. For, as stated on page 307, no causal loop diagram is ever 'finished'. As we learn more, as our understanding deepens, our causal loop diagrams evolve. And if you do wish to enhance any of my diagrams, please let me know, so that your mental model can enrich mine! You can contact me on dennis@silverbulletmachine.com.

PART 1 Systems

Chapter 1

Systems and Mental Models

DOI: 10.4324/9781003304050-2

A (Very) Short Story…

Sam was having a difficult time at home, and arrived at the office grumpy and irritable.

But Sam needed Alex's support on a particular issue to be discussed at a meeting that morning.

Alex, though, was also feeling down. So Alex interpreted Sam's unease as personal hostility, and was at best lukewarm at the meeting.

That made Sam even grumpier, so much so that Sam dashed off an email… an email Sam would live to regret…

Well, we'll all agree that I'm not going to win a Nobel Prize for Literature, nor am I likely to be commissioned to write a script for a soap opera. But my intention here is not to write a novel; rather just to point out that much of what we all experience can be represented as a 'story', a sequence of events that unfold over time, in which different people do things, make decisions, take actions, influence one another. And we all know that the most compelling stories are about how decisions play out, about what happens when the consequences of that fateful decision to [steal that cash]/[open the door to the stranger]/[send an aggressive email]/… finally crystallise.

Many stories are told in novels; others in film or on television. But there's a particular category of story that is told in none of these ways. In fact, except very rarely, these stories are not told as stories at all, but as dry, lengthy, often hard-to-read texts to be found in the 'management' section of the local book shop or on-line store (the exceptions, by the way, are some types of business school case study).

This book breaks the mould. Into shatters.

For it tells business (and related) stories, and tells them not in words, but in pictures. Pictures known as **causal loop diagrams**; pictures that tame the often overwhelming complexity of many real situations; pictures that describe how complex **systems** behave over time; pictures that draw on that most powerful methodology known as **systems thinking**. And let me mention here that a convention I'm using in this book is to identify particularly important technical terms by using **bold font, in red,** when each is first introduced.

Systems

This book is about the behaviour of systems. Systems are important, for we interact with them all the time, and many of the actions we take are influenced by a system – for example, the system of performance measures in any organisation influences, often very strongly, how individuals within that organisation behave. Furthermore, sometimes we are involved in the design of systems, as is any manager contributing to the definition of what those performance measures might be. That manager will want to ensure that all the proposed performance measures will drive the 'right' behaviours, rather than (inadvertently) encouraging dysfunctional 'game playing', and so anticipating how the performance measurement system will work in practice is a vital part of a wise design process.

Some of the systems with which we interact are local, such as my organisation's performance measurement system. Some systems, however, are 'distant', but none the less very real, such as the healthcare system, the education system, the legal system, the climate system. Systems therefore exist on all scales, from the local to the global. And all systems are complex, some hugely so. That's why understanding how systems behave can be very helpful.

Systems are complex for three main reasons.

First, the manner in which they behave over time can be very hard to anticipate – and anticipating the future sensibly is of course a key objective of management.

Second, the 'entities' within a system can be connected together in very complex ways, so that an intervention 'here' can result in an effect 'there', perhaps a long time afterwards. Sometimes this can be surprising, and so we talk of 'unintended consequences' – but this is of course a euphemism for 'because I didn't understand how this system behaves, I had not anticipated that'.

And third, in all real systems, the principle 'entities' within those systems are human beings, human beings who often have choices as regards which of several possible actions to take under any circumstances, human beings who have feelings such as anger or anxiety, human beings who themselves are very complex.

Systems thinking, the subject matter of this book, is the disciplined study of systems, and causal loop diagrams – the 'pictures' of this 'picture book' – are, as we shall see, a very insightful way to represent the connectedness and behaviours of the entities from which any system is composed, so taming that system's complexity.

A Formal Definition of 'System'

This book defines a system as

'a community of connected entities'.

This emphasises the connectedness between the entities within the system and recognises that it is the connectedness across the entities that might be of more significance than the specific properties of any particular single entity.

This definition is very broad, and includes, for example, the healthcare system, in which the 'entities' are GP surgeries, pharmacies, hospitals… as well as the GPs, the pharmacists, the nurses and the patients too…, all of which need to be very well 'joined up' to deliver the required service. And note that the system is not just the physical entities such as the hospitals, but the people as well.

Any business, any enterprise, can be regarded as a system, and for the business, or the police force, to be successful, the various entities within it – for example, the marketing function and the production function, or the different branches of the police service – need to communicate and act purposefully together, rather than each 'doing their own thing'.

Emergence

I went to the bank.

That appears to be a rather dull sentence. But if the definition of a system is 'a community of connected entities', then this is indeed a system, a system in which the entities are words in the English language, and the 'community' is the sentence.

But in fact, it's not dull at all, for it embraces some 'magic' in that this system has an intriguing property.

Meaning.

Everyone reading this book knows what that sentence means, everyone can picture in their mind what is happening.

What makes this sentence 'magical' is revealed by the question 'where, precisely, is that meaning located?'

To which the answer cannot be that the meaning is attributable to any single word. You could study the word 'went' for a lifetime, yet not discover the meaning of 'I went to the bank'. Rather, the meaning is a property of the entire sentence, taken as a complete whole. Meaning is an attribute of the system, not of any of the entities from which that system is comprised.

That's not all. Here are three other 'systems' formed from English words: 'I went to the', 'I went to the bank trousers' and 'The went to I bank'. By comparison to 'I went to the bank', the first has something missing; the second has too much; and the third has the same words but in a different order. And all three lack that magical property of meaning.

Meaning therefore does not result from any arbitrary sequence of words – rather, the right words (not too few, not too many) have to be sequenced in the right order. The 'system' has to be 'just right'. And when it is, meaning emerges.

That's an example of a particular feature of a well-constructed, 'just right', system, a feature known technically as **emergence** – the appearance of a property at the level of a system-as-a-whole, a property that does not exist at the level of an individual entity, or subset of entities, from which that system is comprised. A good system will often show beneficial emergent properties, in that it will work well, whereas a poorly designed system won't. A key purpose of this book is therefore to provide guidance as to how to design good systems, and how to avoid designing bad ones.

Feedback

I went to the bank.

What happened next?

You might like to think about that for a moment.

If your response was something like 'and withdrew some money', 'made a payment', or 'enquired about my account', you are not alone. When I ask that question at my workshops, most people reply with something related to money.

But what actually happened was this:

I went to the bank, sat down on the grass, and watched the swans glide past.

That rather longer sentence is still a system, a 'community of connected entities', a system in which the component parts – some words in the English language – are connected together in just the right way to result in the emergent property of meaning. And that meaning shows quite clearly that the 'bank' is not the building in which financial transactions are conducted, but rather the bank of a river.

The word 'bank' is ambiguous: it can mean 'financial building', it can mean 'land alongside a river'. The simple sentence 'I went to the bank' can refer to either, and most people assume the building. It's not until we get more information, about the grass, about the swans, that it becomes clear that it's not the building but the riverside. That later information refers back to the word 'bank', clarifying its meaning, and resolving the ambiguity. Technically, that's known as **feedback**, and, as we shall see throughout this book, feedback is a central feature of very many systems. The systems we shall be examining, however, won't be linguistic systems composed of words; rather, they will be systems relating to managerial and organisation life in which the connected entities are tangible, measurable, items such as 'sales volume', 'price', 'profit' and 'budget' as well as intangible, but none the less real, and important, items such as 'motivation', 'ability to cope' and 'grumpiness'.

Mental Models

'No,' said Sam. 'Spending on advertising is wrong. We should use the money to expand our distribution. We're right on the edge and we'll have difficulty meeting any further demand – especially as might happen with a successful advertising campaign.'

'No,' replied Alex. 'Advertising will increase sales and boost profits. That's good news. Expanding our distribution is just tipping money down the drain. Our current distribution is just fine.'

Yes, we've all been at that meeting – and we've all been 'Sam' or 'Alex'. And the issue that Sam needed Alex's support on – as mentioned in the story on page 4 – was to invest in distribution, not advertising. But that was not to be.

Neither Sam nor Alex is 'right' or 'wrong'; what's happening here is a demonstration that Sam and Alex have different views on how best to build their business, how best to invest a particular sum of money. Alex wants to boost sales, which is not surprising given that Alex is the Sales Director; Sam is concerned that the distribution capability might become over-stretched, which, once again, is what you might expect of the Distribution Director.

The fundamental question, of course, is 'What is the most beneficial way of investing [this amount] of money?'. And that's a complex question, for the business is a complex entity, operating in a complex competitive market. And, understandably, Alex has one 'view of the world', from the prospective of marketing, whilst Sam has a different view, from the perspective of distribution. In the jargon of systems thinking, Alex and Sam have different **mental models** of how that complex system, their business, operates.

But if they shared the same mental model, if they had the same view of what might happen 'there' as the result of an intervention – such as an investment – 'here', then these arguments wouldn't happen. Which is why this book is important. For the key technique – the causal loop diagram – allows 'me' to define my mental model of any given system very clearly, and to show it to 'you'. I can then say, 'This is how I think the 'world' works – what do you think?' And in the resulting discussion, we can converge on a shared view. Which is a powerful thing to do. And not just in business. Much political conflict is driven by (significantly) different mental models, and politicians and policy makers could derive much benefit from this book too!!!

Purpose, Outcomes and System Design

An important feature of my definition of 'system' – 'a community of connected entities' – is something that's missing.

This definition makes no reference to 'purpose'.

That might be surprising, for isn't one of the most important aspects of a system its purpose? The performance management system within any organisation, for example, exists so that the organisation can indeed do just that, manage its performance, with the various performance measures set to ensure that all departments and managers meet their respective targets so that the enterprise as a whole stays on budget. Likewise, the purpose of the healthcare system in any country is to deliver effective healthcare to the country's population. Shouldn't 'purpose' therefore be included within the definition of 'system'?

The omission of 'purpose' from my definition, however, is intentional. For no system, in and of itself, has a 'purpose'. Rather, a system delivers outcomes – outcomes that depend on the system's structure, and on the behaviours of the people who interact with it.

In so far as there is an associated purpose, that purpose is the intent of those who designed the system – not of the system itself. The extent to which the actual outcomes of the system's operation in practice do or do not comply with the originally-intended purpose is therefore an indication of how well, or poorly, the system was designed and implemented. This is especially the case for systems intended to exhibit emergent properties such as 'better healthcare', for, as discussed on pages 7 and 8, this requires that all the 'right' system components are connected in exactly the 'right' way, with all the feedback sending just the 'right' signals at the 'right' times.

Accordingly, if a system, as implemented, is to deliver outcomes that are in accordance with the designer's original purpose, it is essential that the system be correspondingly well-designed and implemented. Which is difficult. As evidenced, for example, by the frequency of system failure, and the phrases we use when this happens – phrases such as 'gaming the system' and 'unintended consequences'. But people can only 'play games' if the system design lets them; and to me, there are no so-called 'unintended consequences' – but there certainly is abundant, and compelling, evidence of woefully poor systems design and even poorer thinking.

Systems deliver outcomes; they don't have a 'purpose'. And for a system to deliver outcomes in line with the designer's purpose, it must be well-designed, which is what much of this book is all about.

Chapter 2

Links and Dangles

DOI: 10.4324/9781003304050-3

The 'Sam – Alex' System

Most people reading the story of Sam and Alex on page 4 are highly unlikely to think, 'Aha! That's a story about the Sam-Alex system!'

But – at least in the story – Sam and Alex 'exist', and so are 'entities', and they are also connected: indeed, the whole story is about their mutual connectedness, and the tensions that emerge as a result.

Given our definition of a system is 'a community of connected entities', Sam and Alex therefore constitute the 'Sam-Alex system'.

Accordingly, the interaction between Sam and Alex is a manifestation of a particular behaviour of that system, and the possibility arises that this behaviour can be interpreted using systems thinking, and described by an appropriate causal loop diagram.

To do this, we need to identify the key features of the system. These can be inferred from our knowledge of the system, knowledge which in this particular case is based on the story. Different people might identify different features: my suggestions are features such as 'Sam's grumpy feelings' and 'Sam's dysfunctional behaviour'.

Although Sam's grumpy feelings are never actually measured, it is in principle possible to associate Sam's grumpy feelings with a number, such that a low number, say, 2, might mean 'mildly irritable', and a higher number, say, 7, might mean 'very grumpy indeed'. Since the 'degree of Sam's grumpiness' can vary, the feature 'Sam's grumpy feelings' is referred to as a **variable**.

In general, any particular system will be associated with a number of different variables. These variables, however, are not just 'all over the place'. On the contrary, they are connected together in very specific ways. The purpose of a causal loop diagram is therefore to depict, very clearly and precisely, just how those variables are connected, for by doing so, we will then be able to understand how the corresponding system actually behaves…

Direct Links

Sam's grumpy feelings → Sam's dysfunctional behaviour

This diagram captures two ideas, both relating to the connectedness between two features of the 'Sam-Alex system' – *Sam's grumpy feelings* and *Sam's dysfunctional behaviour*. Note that a convention to be used in this book is to use *italics* when referring to any variable explicitly shown in a causal loop diagram.

The first idea relates to causes and effects. According to the story, *Sam's dysfunctional behaviour* on arrival at the office is the result of Sam's *grumpy feelings*, rather than the other way around: Sam did not become *dysfunctional* because of something that happened after he entered the office, he was *dysfunctional* because he was already *grumpy*. The diagram is therefore a visual representation that *Sam's grumpy feelings* cause Sam's *dysfunctional behaviour*, with the cause, *Sam's grumpy feelings* being at the 'blunt' end of the connecting arrow, and the effect, Sam's *dysfunctional behaviour*, at the 'pointy' end. The arrow is known as a **causal link**, or just **link**.

The second idea relates to directionality, in that the *grumpier* Sam becomes, the more *dysfunctional Sam's behaviour*: the variable at the 'blunt' end of the linking arrow (the cause) and the variable at the 'pointy' end (the effect) are both moving in the same direction, increasing together. Furthermore, this remains true for quite large changes in the magnitudes of the variables – if Sam is already really *grumpy*, and gets even *grumpier*, the already highly *dysfunctional behaviour* is likely to become even worse.

If an increase in a cause variable drives an increase in an associated effect variable, implying that the two variables move in the same direction, the corresponding link between them is known as a direct, or positive, link, as represented in this book by a solid arrow.

Direct Links (Usually) Work 'Both Ways'

Sam's grumpy feelings → Sam's dysfunctional behaviour

On the previous page, we saw that when *Sam becomes more grumpy*, his *behaviour becomes even more dysfunctional*, as represented by the direct link, with the variables at each end of the linking arrow moving in the same direction, upwards.

Conversely, the less *grumpy Sam's feelings* (this being a decrease in grumpiness), the less *dysfunctional Sam's behaviour* (this being a decrease in dysfunctionality). The two variables at each end of the linking arrow are once again moving in the same direction together, but this time both downwards.

In this example – which is one of very many – the direct link works 'both ways' with an increase in the variable at the 'blunt' end of the arrow driving an increase in the variable at the 'pointy' end, and also a decrease driving a decrease.

There are, however, certain situations – as will be discussed on page 56 – in which a link acts in only one direction, in that, for these 'one-way links', the 'both downwards' effect is not observed. The 'both upwards' effect, however, is universal, and so the definition of a direct link, as given on the previous page – that an increase in the 'cause' variable drives an increase in the associated 'effect' variable – is always true.

Finally in this discussion of direct links, let me note some other sources represent direct links differently, identifying them by a + sign, or by the letter ***S*** (for 'same'), towards the 'pointy' end of the arrow, as shown here:

Continuing the Story...

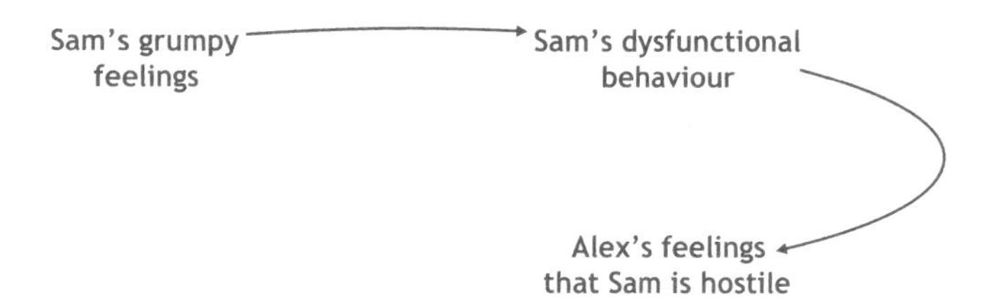

Alex notices *Sam's dysfunctional behaviour* and interprets this as *hostility*, and so, as the cause, *Sam's dysfunctional behaviour*, is at the 'blunt' of the linking arrow to the effect, *Alex's feelings that Sam is hostile*, at the 'pointy' end.

Also, the more explicit *Sam's dysfunctional behaviour*, the stronger *Alex's feelings that Sam is hostile*. The connecting link is therefore a direct link, as represented by the solid arrow. And it works in the other direction too: the less *dysfunctional Sam's behaviour*, the weaker *Alex's feelings that Sam is hostile*.

...One Step Further...

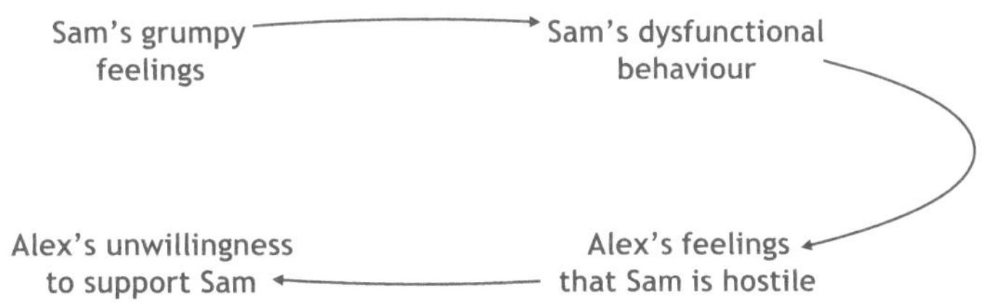

Sam's grumpy feelings → Sam's dysfunctional behaviour → Alex's feelings that Sam is hostile → Alex's unwillingness to support Sam

Alex's feelings that Sam is hostile then drive Alex's unwillingness to support Sam...

This has been represented as a direct link.

Does that make sense?

Do you agree?

A Feedback Loop...

And *Alex's unwillingness to support Sam* makes *Sam even grumpier.*

As you may verify if you wish, this diagram captures all the elements of the story, as told in words, on page 4. But very succinctly.

Also, as can at once be seen, the diagram now forms a closed loop, known as a **feedback loop**, which is both important and informative, for although the diagram is inherently static, we can infer how the 'Sam-Alex system' will behave over time, as shown on the next page.

...Which Behaves as a (Very) Nasty Vicious Circle...

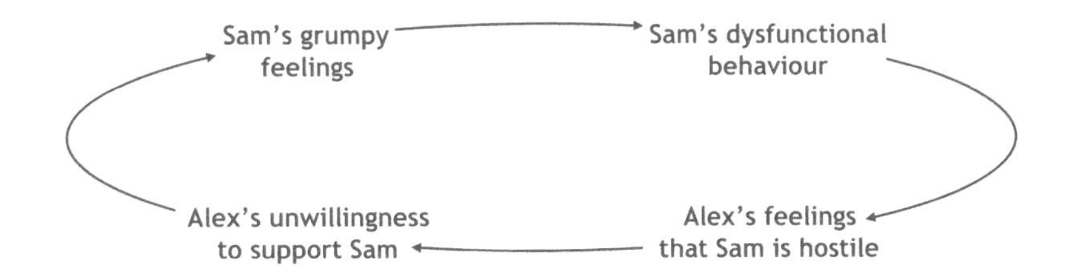

Alex's unwillingness to support Sam at the meeting makes *Sam gets even grumpier...*

...increasing *Sam's dysfunctional behaviour*, for example, as in the story, by sending an aggressive email...

...so fuelling *Alex's feelings that Sam is hostile...*

...which in turn makes *Alex even more unwilling to support Sam...*

...causing *Sam to become grumpier still...*

...increasing *Sam's dysfunctional behaviour* even more...

That really is a very nasty 'vicious circle' which gets progressively worse on each turn.

The fact that the various features of the loop are reinforced on each cycle gives this type of feedback loop the technical name **reinforcing loop** or **positive loop**.

...The Short Story, Continued

Chris, a friend of Sam, noticed what was happening at the meeting, and was uncomfortable observing the increasing rift between Sam and Alex.

'Hi Sam', Chris said, later that day. 'Things don't seem to be going so well between you and Alex, and I did spot that you were pretty grumpy at that meeting this morning. Would you like to talk about things? Is there anything I can do to help?'

At first, Sam was reticent, but Chris was a trusted friend. 'Yes, you're right', said Sam. 'I was feeling down before I came to the office, and unfortunately Alex was at the wrong end of my bad mood. And sending that email afterwards didn't help...'.

'Well, we've all sent emails like that,' replied Chris. 'Why not contact Alex and suggest having a cup of coffee together to talk it all over...'.

'Mmm... thank you... yes... that does make sense...'

Sam might have rejected Chris's interference. But that's not what happened. As a result of the conversation with Chris, Sam agrees to tone things down – so that suggests a causal link between *Chris's constructive intervention* and *Sam's dysfunctional behaviour*. But this link is different from the all the links we had seen so far: as *Chris's constructive intervention* becomes progressively more persuasive, *Sam's dysfunctional behaviour* does not become even more dysfunctional, but rather abates.

Inverse Links

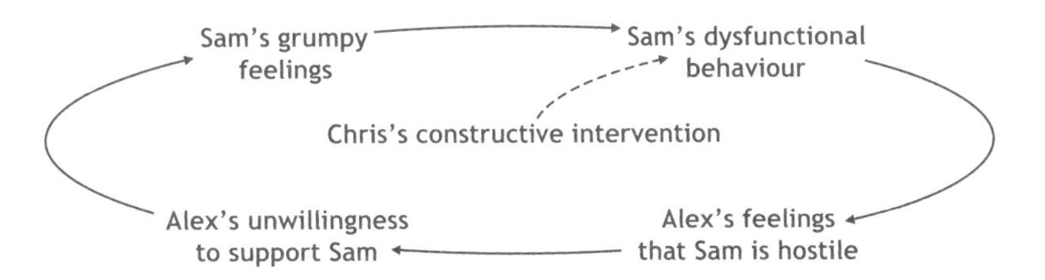

Yes, there is a causal link from *Chris's constructive intervention* to *Sam's dysfunctional behaviour*, and as the variable at the 'blunt' end of the arrow, *Chris's constructive intervention*, the cause, becomes stronger, the variable at the 'pointy' end of the arrow, *Sam's dysfunctional behaviour*, the effect, becomes weaker.

So far, all pairs of linked variables have moved in the same direction, and so have been associated with direct links. This situation is different, for the linked variables are moving in opposite directions. The corresponding link therefore cannot be a direct link; rather, it is a new form of link, an **inverse link**.

If an increase in a cause variable drives a decrease in an associated effect variable, so that the two variables move in opposite directions, the corresponding link between them is known as an inverse, or negative, link, as represented in this book by a dashed arrow.

Some other sources represent an inverse link by a solid arrow associated with a – sign, or the letter ***O*** (for 'opposite'), close to the arrow head, as illustrated here:

Also, as with direct links (see page 14), some inverse links - technically known as 'unidirectional outflows' - act only 'one way' (see pages 56 and 312).

Slowing the Vicious Circle Down

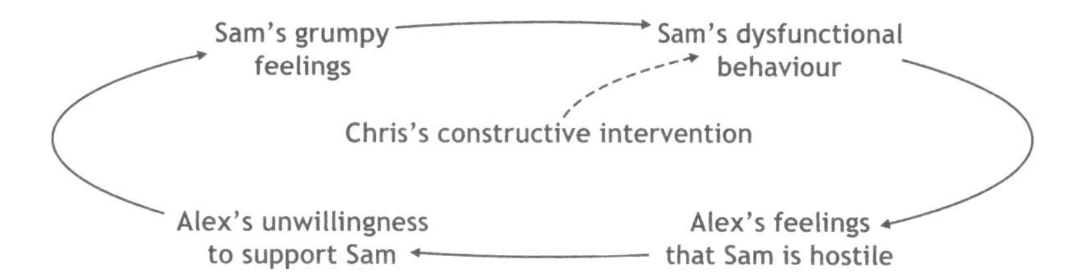

If Chris had done nothing, then this vicious circle would have driven an escalation of tension between Sam and Alex. But *Chris's constructive intervention* tempered *Sam's dysfunctional behaviour*, so that when Sam and Alex met for coffee, *Alex sensed rather less hostility* and so was rather more muted in expressing *unwillingness to support Sam*...

...which in turn caused *Sam to feel rather less grumpy*...

...so slowing the vicious circle down...

...and possibly even stopping it, especially if Sam and Alex can agree to work together.

Dangles

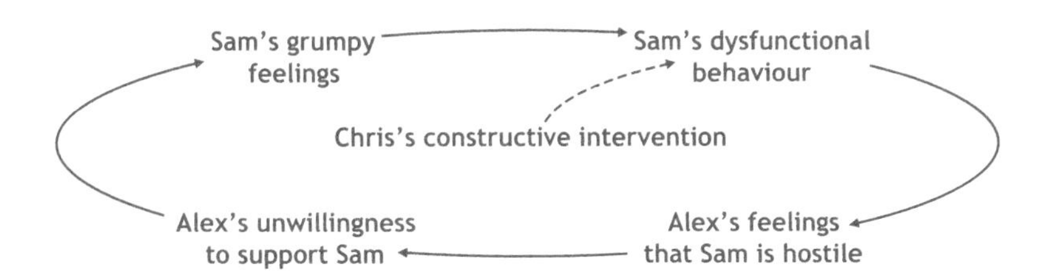

Collectively, the four variables *Sam's grumpy feelings*, *Sam's dysfunctional behaviour*, *Alex's feelings that Sam is hostile* and *Alex's unwillingness to support Sam* constitute the 'vicious circle' reinforcing feedback loop, and they each have the property of being connected 'both ways', in that each variable is simultaneously an effect of a preceding variable, and also a cause of a succeeding variable. This is evident visually since each of these variables has at least one 'incoming' link, as well as at least one 'outgoing' link.

The variable *Chris's constructive intervention* is different in both respects. First, although it is connected to the closed feedback loop, it is not within it; second, it is connected 'only one way' in that it influences, and hence is a (partial) cause of, *Sam's dysfunctional behaviour*, but *Chris's constructive intervention* is not the result of any variable shown. This is evident visually, for there is no 'incoming' link.

Any variable that connects only 'one way' is known as a **dangle**, of which there are two main types. **Input dangles** have only 'outgoing' links, connecting 'into' the diagram, and so are only causes: *Chris's constructive intervention* is an example. **Output dangles** have only 'incoming' links, connecting 'out of' the diagram, and so are only effects.

Although *Chris's constructive intervention* is an input dangle, this does not imply that there are no variables that it might be influenced, or caused, by. There may well be, but they are not shown in the diagram.

Dangles therefore define a system's boundaries, defining what 'comes into' the system from 'outside', and what 'leaves' the system. Accordingly, input dangles drive the system, and the output dangles are the system's outcomes.

Chapter 3

Causal Loop Diagrams

DOI: 10.4324/9781003304050-4

An Example of a Causal Loop Diagram

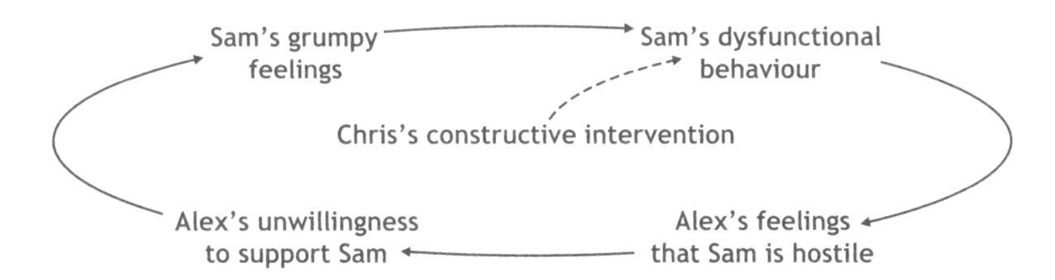

This is an example of a **causal loop diagram**, and the processes associated with the drafting, confirmation and interpretation of causal loop diagrams for any system of interest are collectively known as **systems thinking**.

This particular causal loop diagram contains just one feedback loop, a reinforcing loop, and is quite simple, as is the system it describes. But as will be seen later in this book, causal loop diagrams can be very complex indeed, containing many interconnected feedback loops, for they are depicting correspondingly complex systems – for example the system describing climate change, as shown on page 288.

Study of an insightful causal loop diagram, however, can tame this complexity, so that the dynamic behaviour of even the most complex systems can be understood. And if, for whatever reason, that dynamic behaviour is considered to be unsatisfactory, the causal loop diagram provides a 'thought laboratory', a 'safe environment' for exploring alternative system structures, and for testing ideas so as to avoid 'unintended consequences', and 'quick fixes' that backfire.

The next few pages therefore explore some further features of causal loop diagrams.

Some Important Points about Links

[This] causes → [That] effect

Throughout this book, we will explore many causal loop diagrams, some very complex, and each containing any number of causal links. These links are fundamental to all the diagrams, and so it's important to appreciate what these links do, and do not, represent.

As we have seen, links indicate both causality and directionality. So, in the generalised diagram above, the variable [*This*] at the 'blunt' end of the arrow is the cause of the effect at the 'pointy' end, [*That*]. [*This*] therefore drives [*That*] directly, in that a change in the intensity or value of [*This*] directly causes a corresponding change in the intensity or value of [*That*]. If an increase in [*This*] results in an increase in [*That*], the link is direct, and represented by a solid arrow; if an increase in [*This*] results in a decrease in [*That*], the link is inverse, and represented by a dashed arrow. And these relationships hold over large variations in the magnitudes of the changes in [*This*] and [*That*].

In any diagram, some links are 'stronger' than others, and in principle, the relative 'strength' of the various links might be distinguished by, for example, using thicker or thinner lines for the arrows. In my experience, however, this adds visual clutter, and can be misleading, so in general all links in this book are shown of the same thickness, and so contain no information as regards relative strength. The only exception will be the occasional use of thicker arrows to draw attention to particular features of a particular diagram.

The fact that [*This*] drives [*That*] leads the mathematically inclined to think 'what's the formula?'. Yes, there might be a formula, but the link makes no reference to it. There is no implication that if [*This*], say, doubles, then [*That*] will double too. The link only indicates a generalised 'up-ness' (a direct link) or 'down-ness' (an inverse link), rather than the mathematical detail.

Also, some links operate more quickly than others, and if a link is especially 'slow', this can be indicated by associating *delay* or *lag* with the arrow:

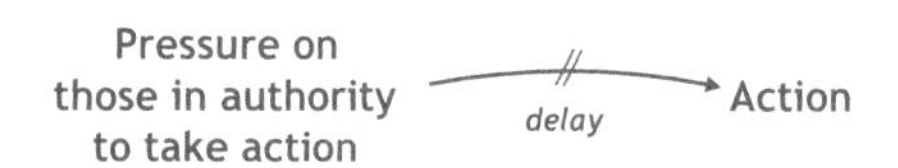

Numbers…

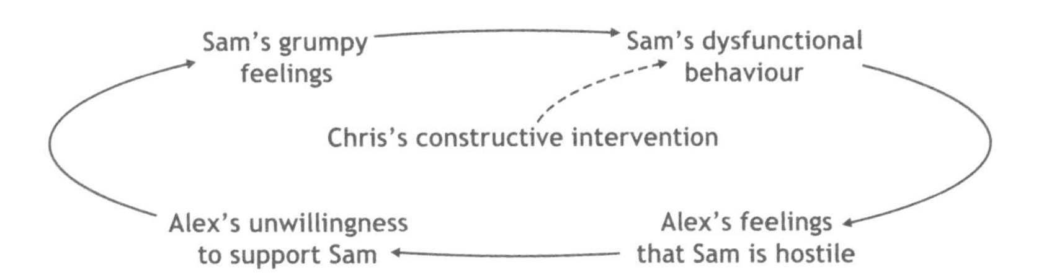

In principle, all the variables in a causal loop diagram can be associated with numbers, and measured. So although measures of *Alex's unwillingness to support Sam* are unlikely to feature in the management accounts of Sam's and Alex's company, we can imagine that a measure of, say, 2 means 'mildly unwilling, but not so unwilling as to make a big fuss' and of, say, 10, 'very unwilling indeed, and will make as much fuss as I can'. Accordingly, as *Sam's behaviour becomes increasingly dysfunctional*, *Alex's feelings that Sam is hostile* become stronger, causing Alex to become progressively more unwilling, so increasing the corresponding measure of *Alex's unwillingness* from, say, 2 towards, say, 5.

Suppose, however, that, as a result of *Chris's constructive intervention*, Sam agrees to talk to Alex, implying that *Sam's dysfunctional behaviour* decreases, driving a corresponding decrease in *Alex's feelings that Sam is hostile*. This in turn causes a decrease in *Alex's unwillingness to support Sam*, implying that the corresponding measure will reduce from, say, 5 to, say, 2.

Suppose further that matters become steadily better, and that *Alex's unwillingness to support Sam*, reduces from 2, to 1, to zero, and even better still, reducing the measure of *Alex's unwillingness to support Sam* even further, from zero to −1, to −2…

...Can Be Negative as Well as Positive...

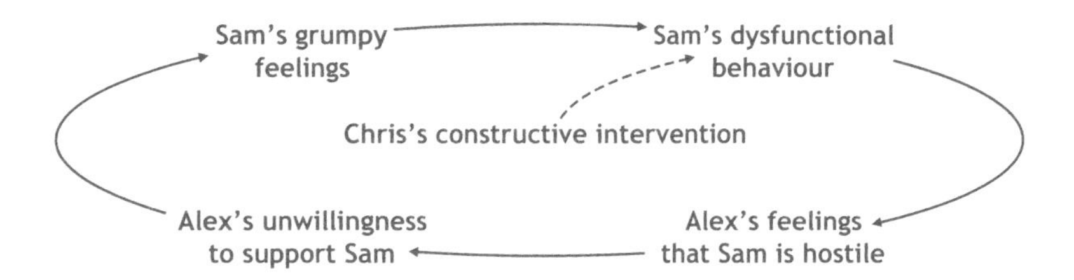

Arithmetically, numbers can be negative as well as positive. But what might a negative number for the measure of *Alex's unwillingness to support Sam* mean in the 'real world', if it means anything at all?

That's worth thinking about. And it does have an answer. Negative 'unwillingness' is positive 'willingness' – so if a value of +2 for *Alex's unwillingness to support Sam* means 'mildly unwilling, but not so unwilling as to make a big fuss', then 0 might mean 'neutral', –2 might mean 'lukewarm'; –5, 'keen'; and –10 'very active advocate'.

This diagram therefore caters for both 'unwillingness' and 'willingness' if we accept that the number associated with this variable can be both positive and negative, recognising that, in everyday language, we use different words.

Many variables that appear in causal loop diagrams might be associated with positive or negative numbers, depending on the context, some common ones being joiners/leavers (as regards an organisation's staff establishment), profit/loss, confidence/stress. Some variables, however, can be associated only with a positive number – for example, a country's population can never be a negative number.

...As Explicitly Shown Here...

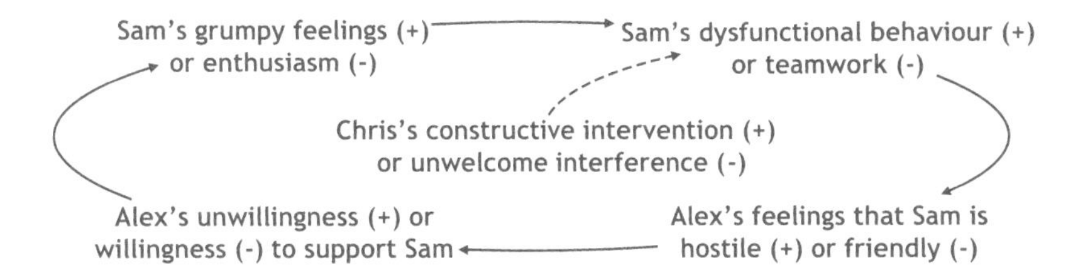

As can be seen from this causal loop diagram, all the variables in the original diagram can be associated with numbers that may be either positive or negative, with different words used accordingly. This diagram, however, is rather cluttered, and so the diagrams presented from here onwards will be more succinct, in the confidence that you will fully appreciate the subtleties of language, and the possibility that some or all of the variables might be associated with positive or negative numbers.

Two further points about numbers. First, it's important not to confuse a 'negative variable' with a 'negative number' associated with that variable. In this example, *Sam's grumpy feelings* is in essence a negative emotion, yet it is associated with a positive number; the converse, Sam's enthusiasm, a positive emotion, is associated with a negative number. In this book, the words 'positive' and 'negative' will be used only in connection with numbers, describing whether they are associated with a + sign or a − sign.

And second, it's important to recognise that the negative number –5 is smaller than the negative number –2. This can be confusing, for the positive equivalents are the other way around: +5 is certainly greater than +2. With reference to the 'number line', with 0 at the centre, and the positive numbers +1, +2, +3... extending to the right, then the negative numbers, –1, –2, –3... extend to the left. So any number [this] to the left of any other number [that] is the smaller number, and the number to the right is the larger number.

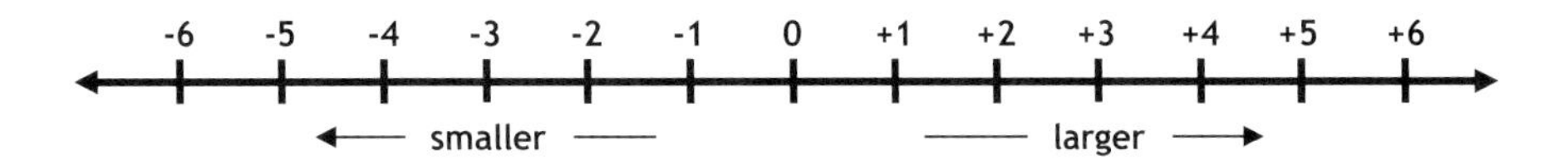

Chris's Intervention

In the diagrams presented so far, Chris's conversation with Sam has been represented by the variable designated *Chris's constructive intervention*, which, because it was indeed constructive, acted to reduce *Sam's dysfunctional behaviour*, implying that the corresponding link is inverse. The alternative possibility that Chris's conversation might have been regarded by Sam as offensive is recognised in the diagram on the previous page by the designation *Chris's constructive intervention* (+) *or unwelcome interference* (–).

But suppose I chose to use the designation *Chris's intervention*, without the associated *constructive* or *unwelcome* descriptions, and so with no indication of the nature of that *intervention*, and how it might be interpreted by Sam.

There surely is a causal link from *Chris's intervention* to *Sam's dysfunctional behaviour*, but is it a direct link or an inverse link?

The issue raised by this question is that *Chris's intervention* can work both ways. If *Chris's intervention* is constructive, then this serves to reduce *Sam's dysfunctional behaviour*, and so the associated link is inverse. But if *Chris's intervention* is regarded by Sam as offensive, this could well accentuate *Sam's dysfunctional behaviour*, in which case the link is direct.

But there can be only one link between any pair of variables, and that link must be either direct or inverse: it can't be both.

There are two resolutions to this dilemma. The first is shown here, where two new variables are introduced, the tempering one acting to reduce *Sam's dysfunctional behaviour*, the aggravating one to increase Sam's *dysfunctional behaviour*. Both cases are covered, and the alternative paths are invoked according to context.

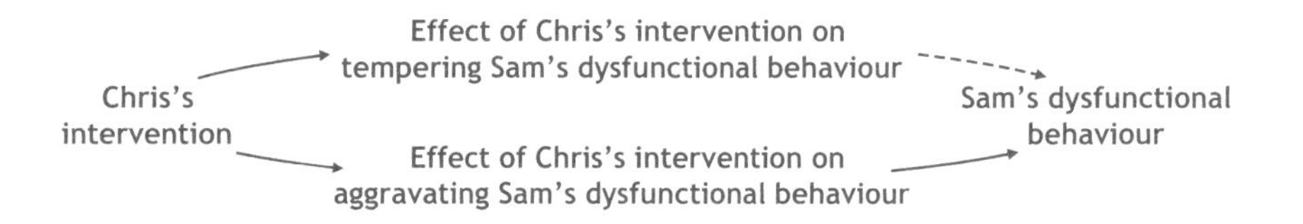

Influence Links

Variables with designations such as the effect of *Chris's intervention on tempering Sam's dysfunctional behaviour*, however, are clumsy, and clutter the diagrams, so there is an alternative representation – the **influence link**:

As shown here, an influence link is represented in this book by a solid arrow with a 'blob' towards the head, and it means, in this example, '*Chris's intervention* has an influence on *Sam's dysfunctional behaviour*, but that *influence* can act in either direction, either increasing, or decreasing *Sam's dysfunctional behaviour*, depending on the context'. Influence links, however, come with a 'health warning':

As a matter of good practice, influence links should ONLY be used in association with dangles, and certainly must NEVER be used within any closed loop.

The restriction on using influence links only in association with dangles – such as, in this example, the input dangle *Chris's intervention* – is very important, for one of the most valuable aspects of systems thinking is that it forces rigour in determining the nature of any particular link as either direct or inverse, the significance of which we shall discuss on pages 76 and 77. That said, influence links are often useful in describing policy decisions that determine input dangles, as illustrated by these two examples:

The Price-Revenue Puzzle

'I'm working on a causal loop diagram, and have got a bit stuck.'

'What's the problem?'

'Well, price and revenue are clearly causally linked. But when you put the price up, sometimes the revenue goes up, sometimes down. So the link between revenue and price must be an influence link, mustn't it?'

No.

Rather, it's an opportunity to think harder.

Yes, it is true that when a *price* is increased, sometimes the *revenue* goes up, sometimes down. Yes, there is a link, but this fact does not imply that the link is an influence link.

In fact it's a direct link, for the very good reason that *revenue* is defined as the result of multiplying the *price* by the *volume* of product or services sold at that *price*. So, by the rules of arithmetic, when the *price* increases (for any constant *volume*), the *revenue* must go up too – so that's a direct link. Similarly, when the *volume* increases (for any constant *price*), the *revenue* also goes up, so that's a second direct link from *volume* to *revenue*.

But that's not the whole story. For most products and services (but not all – some luxury goods behave differently), an increase in *price* drives a decrease in *volume*, corresponding to an inverse link.

Accordingly, the relationships between *price*, *volume* and *revenue* may therefore be represented as:

This shows that there are two paths from *price* to *revenue*. If the *price* is increased, the result of the 'upper' path is to increase the *revenue*, but the result of the 'lower' path is to decrease the *revenue* by virtue of the decrease in *volume*. These two paths operate simultaneously, and the overall outcome depends on the circumstances.

Some Thoughts on Causal Loop Diagrams, and on Links

As we shall see throughout this book, the variables that can appear in causal loop diagrams are of many different types, and are not limited to variables that conventionally appear in a company's management accounts, or are routinely monitored in, say, a corporate 'balanced scorecard'. Rather, they represent whatever is relevant to the matter under study, as, for example, *Sam's grumpy feelings*.

That said, many causal loop diagrams contain familiar variables such as, total costs, profit, revenue and capacity utilisation – indeed, one of the great benefits of systems thinking in general, and causal loop diagrams in particular, is the opportunity to trace the causality, for example, from 'stress' to 'likelihood of error' to 'cost of correcting errors' to 'total costs' and then to 'profit'.

Overall, this connects an 'intangible' concept, stress, to the accounting variable, profit, which is a real connection and an important one too, but one often not examined in this way.

Compiling a good causal loop diagram requires careful thought; careful thought about the variables to be included, and those left out; careful thought about how the various variables are causally related; and – as we saw illustrated on the previous page in the discussion of the 'price-revenue puzzle' – careful thought as regards each individual link.

As regards links, a good rule is

'Any link must be either direct or inverse, with some very few exceptions of influence links, which can be used for dangles only.'

Links and Arithmetic

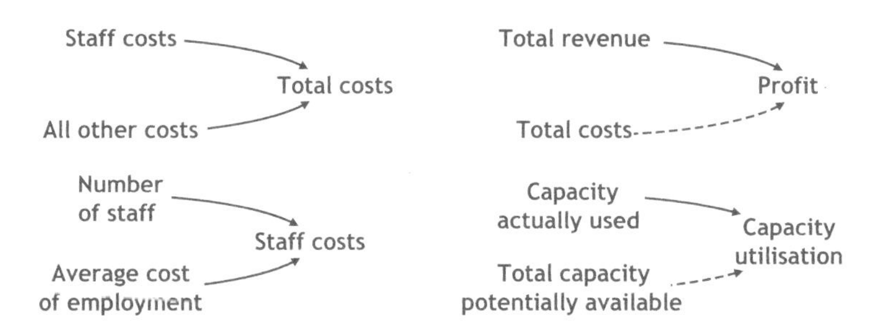

Some variables within any causal loop diagram are defined by arithmetical relationships, for example:

total costs = staff costs + all other costs

profit = total revenue – total costs

staff costs = number of staff × average cost of employment

capacity utilisation = (capacity actually used / total capacity potentially available) × 100

In the context of systems thinking, all the variables on the right-hand side of each of the = signs in these expressions are causes of the corresponding effect on the left-hand side.

As regards the link from any cause to the corresponding effect, and assuming that the number associated with any cause is itself positive:

- For additions, both links are direct.
- For subtractions, the first (positive) variable is associated with a direct link; the second variable, following the – sign, is associated with an inverse link.
- For multiplications, both links are direct.
- For divisions, for the numerator (on top), the associated link is direct; for the denominator (the bottom), the associated link is inverse.

Links and Language

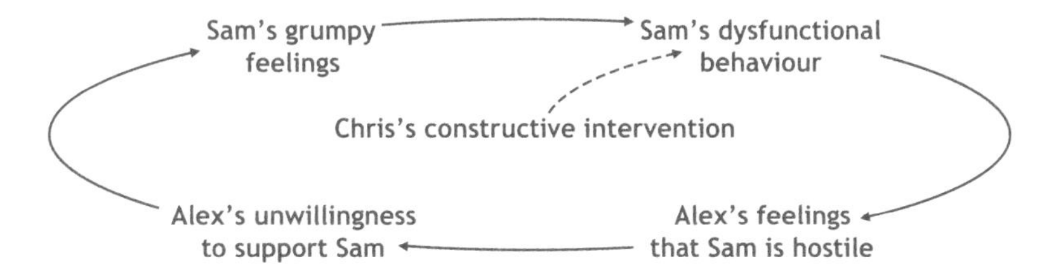

These two diagrams tell the same story – as explained on page 35.

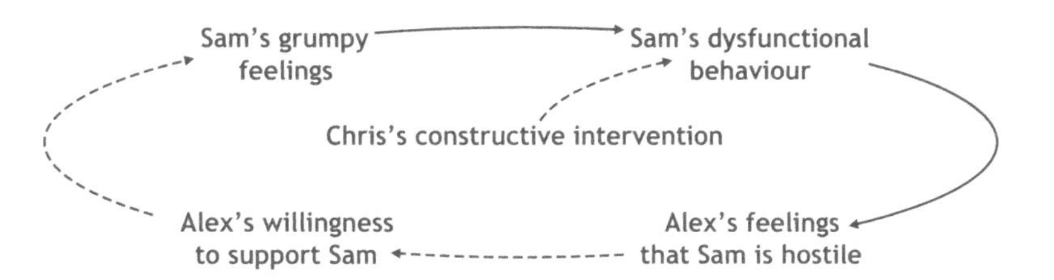

Links and Language

Suppose that, instead of using talking about *Alex's unwillingness to support Sam*, I had used the phrase *Alex's willingness to support Sam*. As usual, this variable can be represented by positive numbers, indicating progressively stronger support, as well as negative numbers, indicating progressively stronger opposition.

Accordingly, as *Alex's feelings that Sam is hostile* increase, *Alex's willingness to support Sam* will decrease, implying that the link from *Alex's feelings that Sam is hostile* to *Alex's willingness to support Sam* is an inverse link.

Similarly, as, *Alex's willingness to support Sam* increases, *Sam becomes progressively less grumpy*, and so that's an inverse link too, completing the lower causal loop diagram on page 34.

Following the lower loop around: as *Sam's grumpy feelings* increase, the direct link implies that *Sam's behaviour becomes progressively more dysfunctional*, and the next direct link implies that this in turn causes an increase in *Alex's feelings that Sam is hostile*. The following link, however, is inverse, and so the increase in *Alex's feelings that Sam is hostile* drives a decrease in *Alex's willingness to support Sam*. According to the next inverse link, *Alex's decreasing willingness to support Sam* results in an increase in *Sam's grumpy feelings*.

The lower loop therefore behaves as a vicious circle in exactly the same way as the upper loop.

But they look different: the named variables are different, and the structure of the links is different too.

Yes. But they describe the same reality, just using different words. That's important, for how causal loop diagrams are presented, and which links are direct and which inverse, will depend on the language used.

Time to Pause...

This is a good place to pause for a moment, for the next three chapters move up a gear. So it's important that you feel fully comfortable with the material presented so far.

As I mentioned on page xxiv in the Preface when I introduced causal loop diagrams, like a map, a causal loop diagram uses a number of standard symbols 'that are readily learnt, as will be explained in the first few chapters of this book'. Those 'first few chapters' are the ones you've just read, and the key symbols are:

- **variables**, which succinctly but clearly identify each item of interest
- **links**, indicating cause-and-effect, represented by 'curly arrows'.

In any causal loop diagram, most variables are associated with at least one link for which the variable under consideration is a cause (and so at the 'blunt' end of a curly arrow), and also at least one link for which the variable under consideration is an effect (and so at the 'pointy' end of a curly arrow). A few variables, however, might only be causes (and so associated only with at least one curly arrow at the 'blunt' end), and are known as **input dangles**, or might only be effects (and so associated only with at least one curly arrow at the 'pointy' end), and are known as **output dangles**. In general, the dangles define the boundaries of the system of interest, with the input dangles being drivers of the system, and the output dangles, the system's outcomes.

Furthermore, links are of two principal types:

- **direct links**, which signify that as the variable at the 'blunt' end of the arrow increases in value or magnitude, so does the variable at the 'pointy' end, as represented by a solid curly arrow
- **inverse links**, which signify that as the variable at the 'blunt' end of the arrow increases in value or magnitude, the variable at the 'pointy' end reduces in value or magnitude, as represented by a dashed curly arrow.

A third type of link, the **influence link**, indicating influence rather than directionality, as represented by a solid arrow with a 'blob' near the 'pointy' end, can be used in association with (usually input) dangles, but only sparingly.

From now on, the material in the rest of the book will use these symbols, without further explanation, to compile causal loop diagrams of progressively greater complexity. So now is an opportunity to look through the previous pages if you wish.

Chapter 4

Reinforcing Loops

DOI: 10.4324/9781003304050-5

The Gerald Ratner Story

At age 15, Gerald Ratner left school without any qualifications, joining the family jewellery business in 1966, and becoming a director when he was 21. For many years, the business languished, and by 1984, there were about 120 shops, making a loss of about £130 M a year. Then Gerald became the boss.

And for the next few years – boom! By 1991, the group owned some 1,500 shops in the UK, and a further 1,000 in the US. Ratner's had become the largest jewelry retailer in the world, and annual profits were running at about £125 M. And the business was valued at about £840 million.

Ratner himself had become 'Mr Retail', and was feted as one of the UK's most successful business figures. And on 23rd April 1991, he was invited to be the guest speaker at an event at London's Royal Albert Hall attended by about 5,000 members of the UK's Institute of Directors.

His speech contained two, now infamous, remarks, presumably intended to be funny.

'People say "how can you sell this for such a low price?" I say because it's total crap.'

'We even sell a pair of gold earrings for under £1, which is cheaper than a prawn sandwich from Marks & Spencer. But I have to say that the sandwich will probably last longer than the earrings.'

Ratner's customers, however, didn't get the jokes.

Rather, they felt insulted. So they stopped buying those earrings.

Within days, the value of the business fell by about £500M to some £340 M; after a few months, to about £14 M, with losses running at an annualised value of about £122 M. Hundreds of stores were closed, and thousands of jobs lost. The Ratner brand name was discarded, and in 1992, Ratner himself was fired.

Bust followed boom. And very quickly too.

Starting a Business...

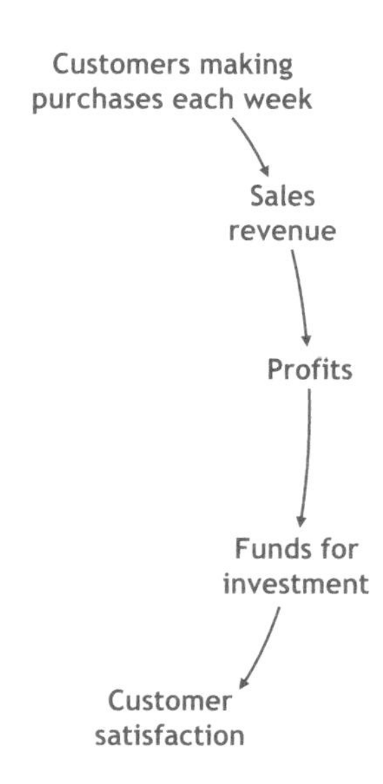

An entrepreneur launches a new business, which attracts some initial customers, who are soon loyal, and make regular *purchases*...

...resulting in some initial, and steady, *sales revenue*...

...which generates some *profits*...

...providing the owner with some *funds to invest* in key aspects of the business, such as marketing and distribution...

...which was money well spent, delivering *customer satisfaction*...

There are of course (very!) many other factors too – factors such as prices, costs (of many different types), interest payments, taxes... all of which could be represented. But they don't have to be, for a valuable feature of systems thinking is that it allows the compiler of a causal loop diagram to be selective. Accordingly, what is, and what is not, included in any one causal loop diagram is a matter of judgement, of choice; judgement and choice as to which variables are essential, important and relevant in any particular context, and which are perhaps just distracting detail, adding clutter rather than insight. And, as we shall see progressively through this book, further items – such as costs and prices – can and will be introduced as they become integral to any particular story.

As a consequence, anyone compiling or using a causal loop diagram must always be alert to the possibility that something important is missing, or that something that has been included might best be deleted. Different people will have different views – and suggestions – that emerge during discussion. Causal loop diagrams are therefore never 'finished', but always evolve and improve. And that applies to the diagrams in this book too – so if you have any ideas and suggestions, please let me know!

...Which Grows...

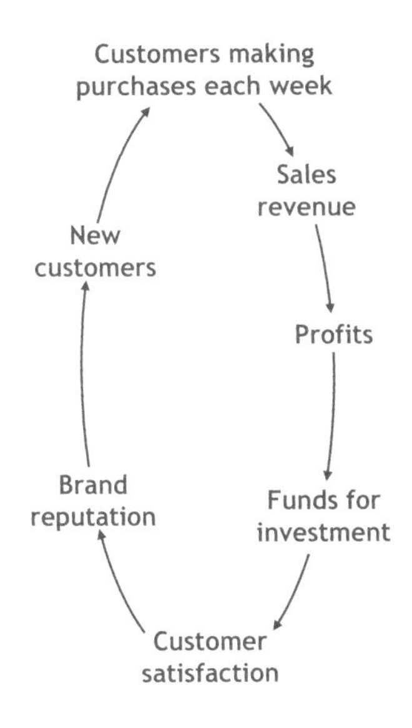

Customer satisfaction can be very effective in building the *brand reputation*, so that not only are existing customers loyal...

...but also *new customers* are attracted...

...adding to the existing *customer base*...

...generating more *sales revenue*.

As a result – and assuming that all costs remain under control as the business grows – *profits* grow too...

...*funding even more investment*...

...so that the *brand strengthens*...

...attracting yet more *customers*...

In every-day language, this is a virtuous circle, driving business growth...

...and in the language of systems thinking, this is a reinforcing feedback loop, with the same structure as the reinforcing loops discussed on pages 17 and 24.

...Exponentially

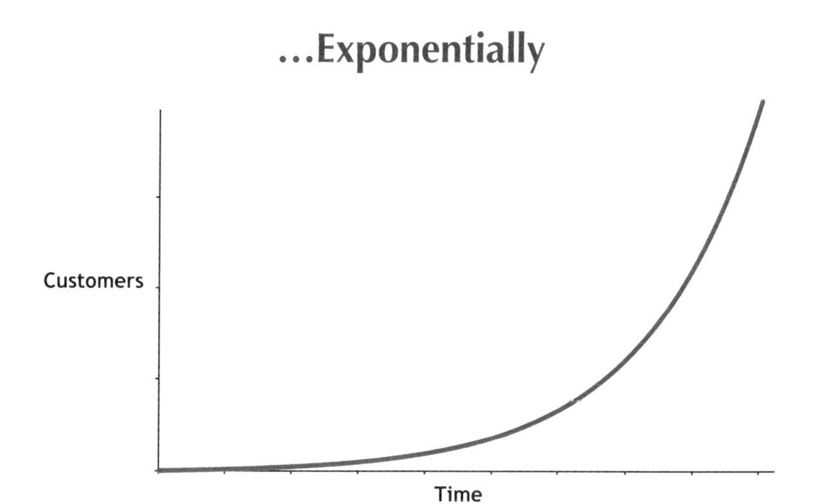

Week	*Customers*	*New Customers*
1	10,000	800
2	10,800	864
3	11,664	933
4	12,597	1,008
5	13,605	1,088
6	14,693	1,175
7	15,868	1,269
8	17,137	1,371
9	18,508	1,481
10	19,989	1,599
...	...	...
20	43,156	3,452
...	...	...
30	93,171	7,454

An important feature of the reinforcing loop of business growth illustrated on the previous page is that it shows that the number of *new customers* attracted to the business is driven by *number of customers already using the business*.

Accordingly, the larger the *number of existing customers*, the larger the *number of new customers*, so making the *number of existing customers* even bigger.

For example, suppose that for every 10,000 *existing customers*, the effect of the *brand reputation* is to attract, say, 800 *new customers* each week. The table to the left shows the *number of existing customers* at the start of each successive week, and also the corresponding *number of new customers*, calculated as 8% of the *number of existing customers*.

As can be seen, the number of customers does not increase linearly; rather, the *number of existing customers* nearly doubles every 10 weeks. This is known as **exponential growth**, at a constant rate of 8% per week.

But Then...

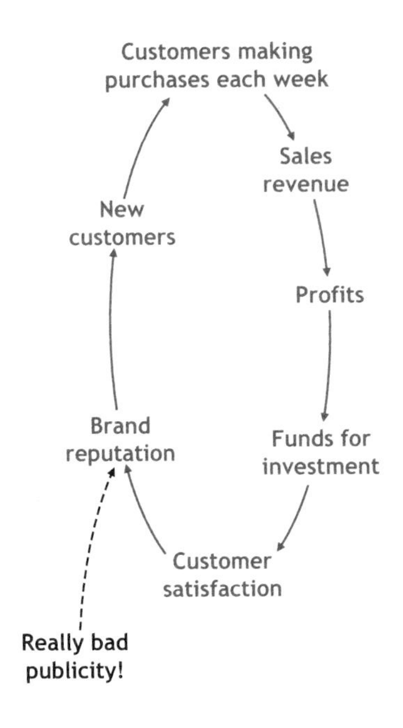

...there was some really *bad publicity* – yes, really bad...

...in fact so bad that the previously positive *brand reputation* becomes significantly damaged...

...and as the *brand reputation* falls, so does the number of *new customers*...

...so much so that soon there were not only no *new customers* at all, but also previously loyal customers no longer make repeat purchases...

...as may be accommodated within this causal loop diagram by allowing the number of *new customers* to become negative, corresponding to the loss of existing customers.

As the *number of customers making purchases each week* plummets, so does the *sales revenue*...

...taking *profits* down too, thereby depleting the *funds available for investment*...

...so those customers who remain are increasingly disappointed...

...and what had been a virtuous circle of exponential business growth suddenly becomes a vicious circle of exponential business decline.

...Bust

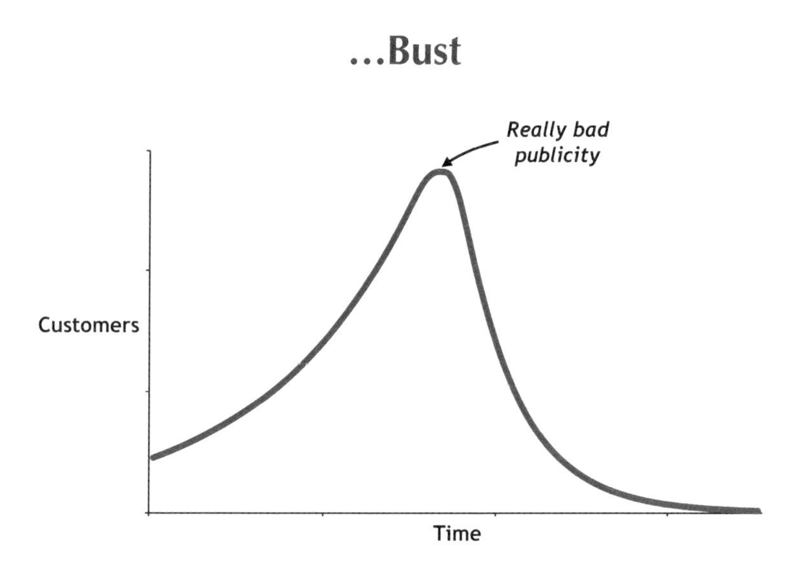

This chart shows the impact of that *bad publicity*, with the initial boom of a virtuous circle suddenly transforming into the bust of a vicious circle – a circle so vicious that the business does not recover. And note that in this illustration the rate of the **exponential decline** is faster than the rate of the original exponential growth.

But as can be seen from the causal loop diagrams on pages 40 and 42, the fundamental structure of the underlying system is the same for both the growth and the decline, with the flip from the one to the other being triggered by the 'shock' of the *bad publicity*. And that fundamental structure is a reinforcing loop, which can drive both exponential growth, as well as exponential decline, depending on how the loop is initiated, and on the effect of any sudden 'shocks'.

And it is because the underlying system is the same that the bust can happen so suddenly, and is so hard to arrest.

Structurally, Vicious and Virtual Circles Are Identical…

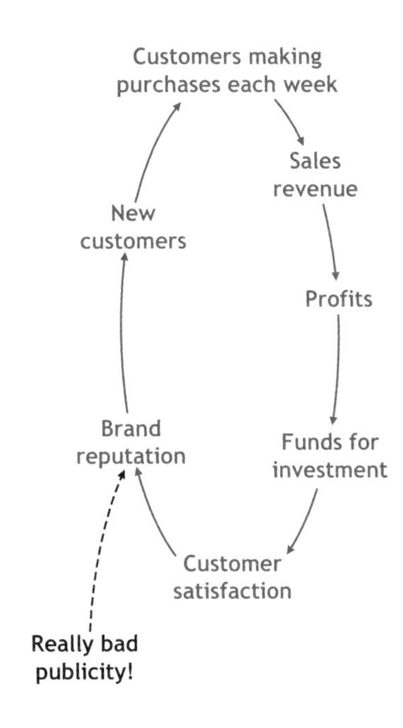

As we saw on page 43, and as was vividly and truly illustrated by the story of Gerald Ratner, this closed reinforcing feedback loop can behave as a virtuous circle driving exponentially increasing business growth, and also as a vicious circle driving exponentially decreasing business decline.

Both behaviours are possible, with everything depending on how the loop is triggered initially, or if, whilst the loop is operating in one way, the system is subject to some form of 'shock', such as *bad publicity*, causing the behaviour to flip from one mode to the other.

Sometimes, a 'shock' is not strong enough to 'flip' the behaviour from growth to decline, but the business is still damaged as the growth rate slows down – the action of the inverse link is therefore to 'put the brakes on'.

...For They Are Just Different Behaviours of the Same Structure...

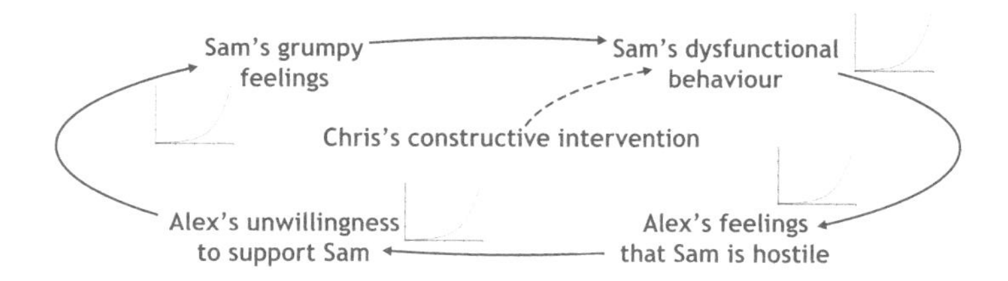

Similarly, the reinforcing feedback loop in the 'Sam-Alex' system can behave in both ways: as a vicious circle if Sam's grumpiness triggers a feud with Alex; as a virtuous circle if the 'shock' of *Chris's constructive intervention* flips *Sam's behaviour* from being dysfunctional (corresponding to positive numbers in this diagram) to friendly (corresponding to negative numbers).

Note that in this diagram, all the variables move in the same direction – depending on how the loop is operating, all increase exponentially, or all decrease exponentially. But that is not always the case...

In the diagram below – which as we saw on pages 34 and 35 behaves in exactly the same way as the upper diagram – as, say *Sam's grumpy feelings* increase exponentially, Alex's willingness to support Sam decreases exponentially, and vice versa.

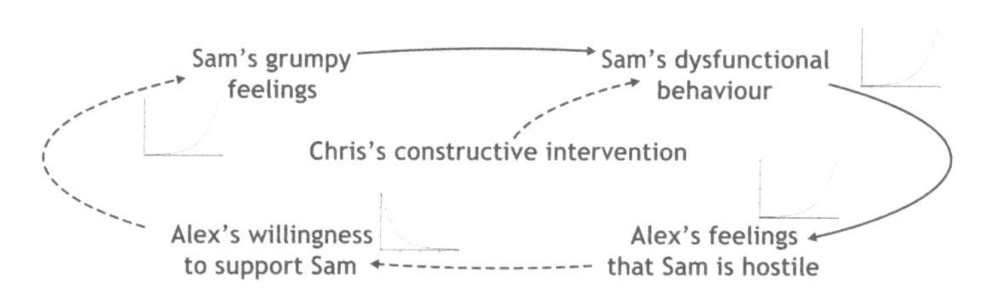

...But Not Every Reinforcing Loop Is Exponential...

A hospital operating theatre can carry out ten non-urgent procedures every day, which in the past has been sufficient, for it was relatively rare that more than ten cases presented on any single day.

Due to the Covid-19 pandemic, however, the hospital had to divert its resources elsewhere, and so was obliged to reduce its non-urgent capacity to two procedures every day, with the result that more and more patients have to wait even longer to be treated...

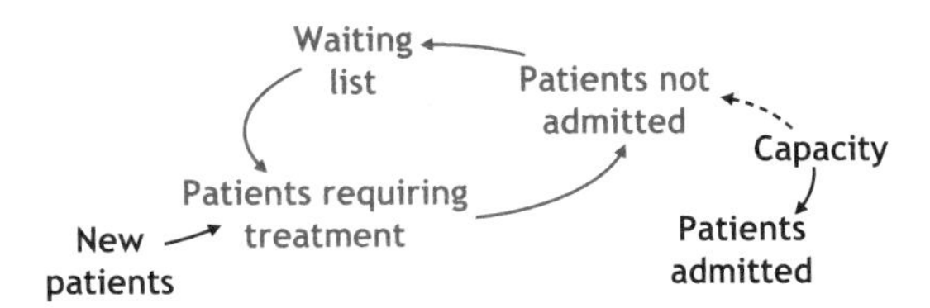

This story can be represented by this causal loop diagram.

Despite the virus, non-urgent *new cases* still arise, and *require treatment*. But since the *number of patients admitted and treated each day* cannot exceed the hospital's *capacity*, the *number of patients not admitted* on any day is the difference between the *number of patients requiring treatment* and the *capacity*.

The *number of patients not admitted* on any day adds to the *waiting list* of patients awaiting treatment, and the *number of patients requiring treatment* is the sum of the *new patients* who present on any day and the current *waiting list*.

The central feature of this causal loop diagram is the reinforcing loop shown in blue, and its operation can be inferred by considering a 'worked example', as shown on the following page.

Note that this causal loop diagram assumes that the *number of patients requiring treatment* is always greater than the *hospital's capacity*, as the story implies. A more general causal loop diagram that does not rely on this assumption is shown on page 111.

...For Linear Behaviour Is Also Possible...

With reference to the causal loop diagram shown on the previous page, suppose that every day, the number of *new patients* who present is 10, and that the *capacity* is 2. The values of the key system variables each day are given in this table:

Day	*New Patients*	*Waiting List from Previous Day*	*Total Number of Patients Requiring Treatment Today*	*Patients Admitted Today = Capacity*	*Patients Not Admitted Today*	*Patients Put into the Waiting List for the Next Day*
1	10	0	10	2	8	8
2	10	8	18	2	16	16
3	10	16	26	2	24	24
4	10	24	34	2	32	32
5	10	32	42	2	40	40
...						
N	10	$8 \times (n - 1)$	$10 + 8 \times (n - 1) = 8n + 2$	2	$8n$	$8n$

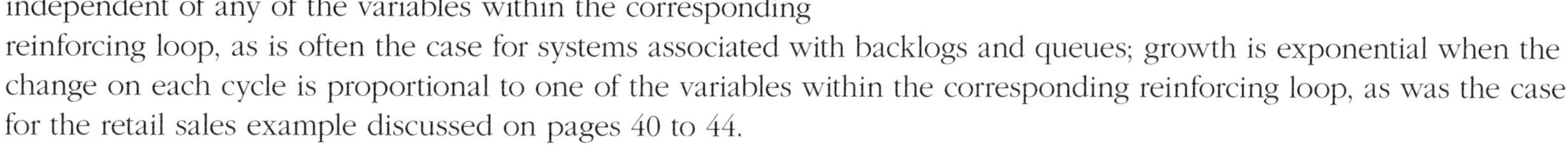

As can be seen from the right-hand column, every day, the *waiting list* grows by 8 patients, 8 being 10 (the number of *new patients* each day) – 2 (the *capacity*). This number is constant over time, implying that the growth in the *waiting list* is linear, rather than exponential.

Although most reinforcing loops show exponential growth or decline, not all do, for, as this example shows, some show linear growth or decline.

Growth is linear when the change on each cycle is constant, independent of any of the variables within the corresponding reinforcing loop, as is often the case for systems associated with backlogs and queues; growth is exponential when the change on each cycle is proportional to one of the variables within the corresponding reinforcing loop, as was the case for the retail sales example discussed on pages 40 to 44.

How to Identify Reinforcing Loops

So far, we have examined three very different contexts: the very personal interaction between Sam and Alex; business growth and decline in general, with a specific reference to the real events concerning Gerald Ratner; and hospital admissions.

Despite the huge differences, all share a common feature – they can all be described by a causal loop diagram structured as a reinforcing loop. Put the other way around, the single structure of the reinforcing loop can explain the behaviour of many, very different, systems. And this is one of the great benefits of systems thinking, for it enables us to identify a deep, single, underlying structure that can explain so many systems which, 'on the surface', appear to be so very different.

So how can a reinforcing loop be identified?

There are some clues on pages 44, 45 and 46, which show four different reinforcing loops. The clues are very hard, and quite non-obvious, to spot 'cold':

A reinforcing loop is any closed feedback loop in which the number of inverse links is even (with zero counting as an even number).

Reinforcing loops show either continuous growth or decline, very often, exponentially.

You can check that on pages 44, 45 and 46: three of the four loops have no inverse links, and the other has two. Zero and two are even numbers, so the loops are reinforcing loops.

This rule is universal. No matter what the closed loop is describing, no matter how complex it is, no matter how many links. If you count the number of inverse links, and if that number is even, say, 0, 2, 4, …, 12…, then the closed loop is a reinforcing loop, and it will always either grow or decline. Always; and very often exponentially.

Chapter 5

Balancing Loops

DOI: 10.4324/9781003304050-6

Pond Weed Grows Really Fast…

A gardener notices a very small quantity of some nasty-looking green stuff growing on a pond.

A colleague has estimated that the entire surface of the pond will be covered in about 20 days' time, and that would be undesirable, for a complete covering will be very bad for the fish.

So the gardener wants to stop that from happening. There is an eco-friendly remedy that can be applied, but it takes 10 days for the appropriate materials to be ordered, and to be delivered.

'The order time is 10 days', thinks the gardener, 'and the pond won't be covered for another 20 days. Since 10 is one-half of 20, I can wait until the pond is half-covered before I place the order.'

Here are some questions:

- ***Is the gardener right?***
- ***On which day is the pond half-covered?***
- ***How much of the pond's surface is covered on the latest day the gardener can safely place an order?***

Exponential Growth Can't Go on For Ever…

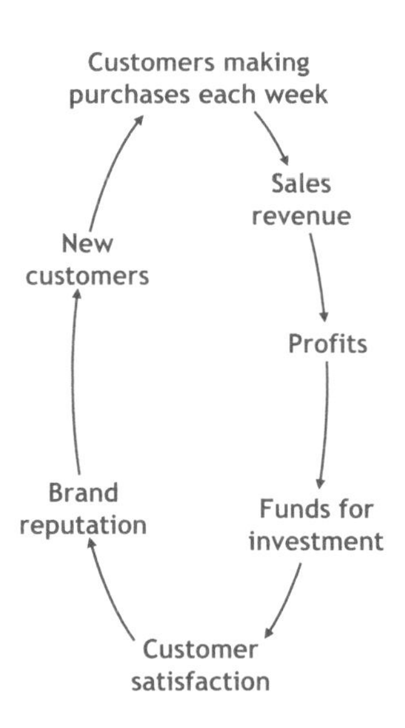

This is the now-familiar reinforcing loop of business growth.

As we have seen, this structure can exhibit exponential growth, with each of the *number of new customers*, the *sales revenue* and the *profits* increasing over time.

According to this diagram, this growth continues for ever, without limit. But in reality, of course, this does not happen. Sooner or later, growth must slow down, implying that this diagram, though 'interesting' and insightful to a degree, is not telling the 'whole story'.

Something must be missing…

...Perhaps Because the Market Has a Finite Size...

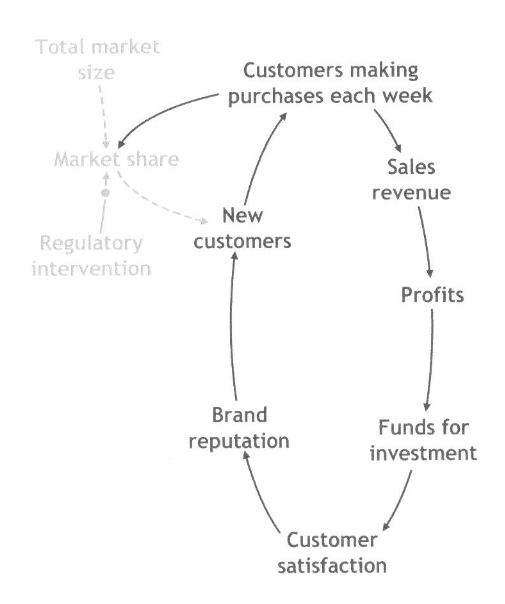

One constraint on business growth might be the *size of the market*, for as the *number of customers making purchases each* week increases, so does the *market* share...

...where the *market share* is the *number of customers making purchases each week* divided by the *total market size* (multiplied by 100 so as to give a percentage)...

...hence the direct link from *customers* to *market share*, and the indirect link from *total market size* to *market share.*

But as the *market share* increases, it becomes increasingly difficult to attract *new customers* – competitors, for example, will have their loyal base, and ultimately there are no *new customers* left once the *whole market* has been captured.

Since the number of *new customers* attracted each week decreases as the *market share* increases, the link from *market share* to *new customers* is inverse.

In addition, a *regulator* may – or may not – act to limit the *market share* of any one supplier. The uncertain nature of this intervention is indicated by the use of an influence link, as shown by the 'blob'.

A Balancing Loop...

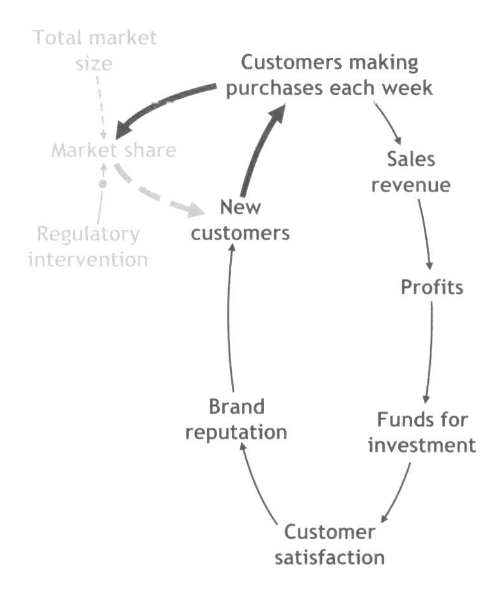

Overall, the progressive saturation of the *market* acts to slow down the rate of growth of the business, ultimately bringing the growth to a halt as the *number of new customers* falls to zero, and the *number of customers making purchases each week* stabilises.

And, as can be seen the variables *customers making purchases each week*, *market share* and *new customers* form a new closed feedback loop...

...but this feedback loop contains just one inverse link, from *market share* to *new customers...*

...for although the total market size and regulatory intervention are associated with the *market share*, the inverse link from *total market size* to *market share* and the influence link from *regulatory intervention* to *market share* are both outside the closed loop, and so are not counted.

As we have seen, any closed loop containing an even number of inverse links must act as a reinforcing loop. But since one is an odd number, the *customers making purchases each week*, *market share*, *new customers* closed loop cannot be a reinforcing loop.

It must be something else.

That 'something else' is called a **balancing loop.**

...Which Acts to Slow Down an Associated Reinforcing Loop...

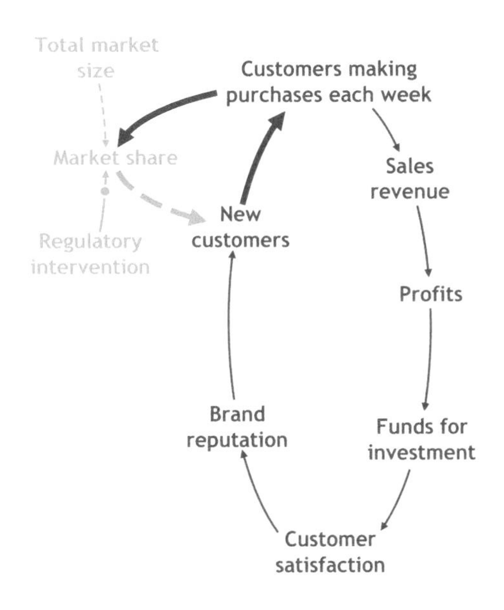

The action of this balancing loop is progressively to slow down the rate of exponential growth associated with the reinforcing loop of business growth, ultimately bring that growth to a halt.

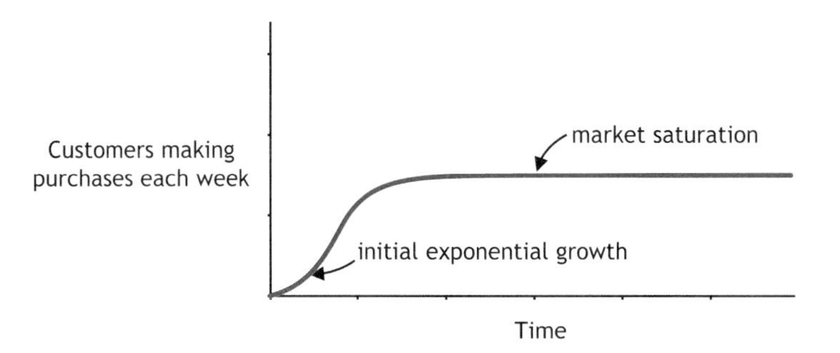

The business solution is to expand into a new market, perhaps by exporting, perhaps by introducing a new product range, so increasing the *total market size* accessed by the business. But, sooner or later, that new *market* will saturate too:

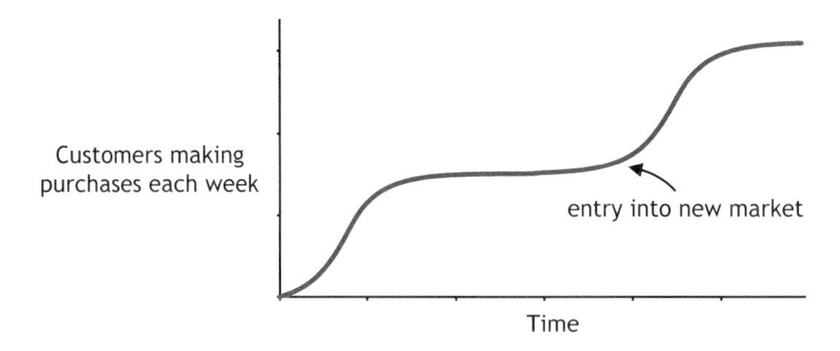

What About Costs?

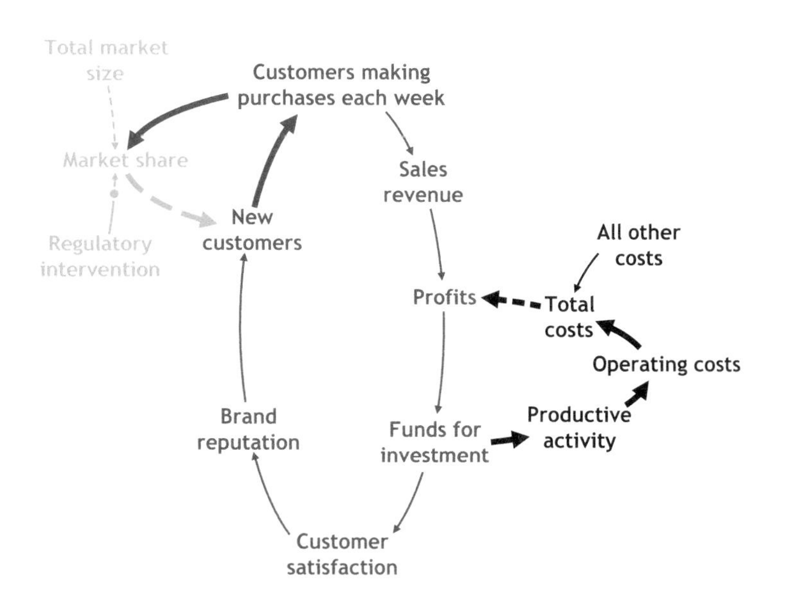

The causal loop diagrams presented so far have made no mention of *cost*. *Costs* are of course of major importance to all enterprises, so this diagram incorporates them in the most general way – noting, as mentioned on page 39, that this diagram is not intended to be an accountant's spreadsheet!

In general, any form of *investment* results in some form of *productive activity* – for example, a retailer or merchant will *invest* in goods to sell on, a manufacturer will purchase raw materials. And everyone will pay staff, all of which contribute to the *costs of operating the activity*.

Other costs, such as rents and interest payments, will also be incurred, and the resulting *total costs* are then subtracted from the *sales revenue* to determine the organisation's *profits*.

The incorporation of *costs* has added another feedback loop, which also has a single inverse link, and so is a balancing loop too. Its acts to make the *profits* less than they would have been otherwise, so further slowing down the reinforcing loop of business growth.

This introduces the possibility that, for whatever reason, the *total costs* might exceed the *sales revenue*, causing the *profits* to become negative – or, in more every-day language, the business might begin to incur *losses*. If these *losses* are of sufficient magnitude, and sustained over time, then the reinforcing loop can flip from exponential growth into exponential decline, and bust will follow boom.

Population Dynamics...

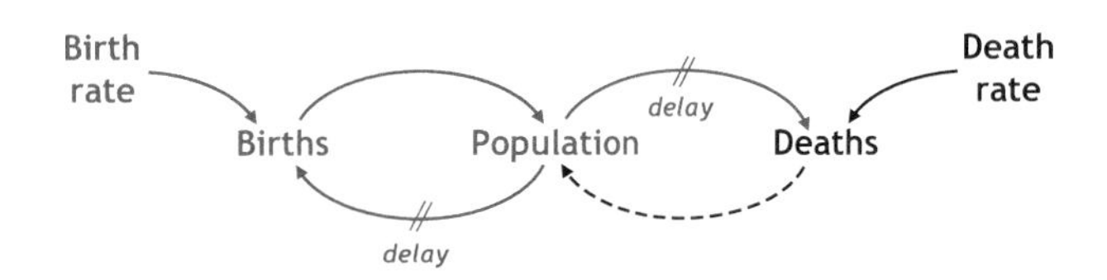

Here is another example of the interaction of a reinforcing loop and a balancing loop, but in a very different context – the dynamics of the population of any living species.

For every living species, the current *population* drives the number of *births*, as determined by the *birth rate*, the number of live *births* for every, say, 1,000 members of the *population* over any defined time period, for example, over any 1 year. For some species, there is a *delay* corresponding for the time to reach adulthood.

Every *birth* adds to the *population*, so completing a reinforcing loop, implying that every living species will increase its *population* exponentially – like the pond weed in the story on page 50 – in principle without limit.

In reality, that does not happen, for as the *population* increases, driving the number of *births*, the number of *deaths* increases too – once again, for most species, with a *delay* corresponding to ageing. The number of *deaths* is determined by the *death rate*, the number of *deaths* for every, say, 1,000 members of the *population* over the same time period as defined for the *birth rate*.

This introduces a second loop, a balancing loop, which acts synchronously with the reinforcing loop, holding back its exponential growth – just like the *market share* balancing loop slowed down the reinforcing loop of business growth.

Note that all the variables in this causal loop diagram are inherently positive; none can ever be negative. Also, the *birth rate* and the *death rate* are each a particular category of input dangle known as a **rate dangle**.

One further point about the link from *births* to *population*. If at any time the *population* is, say, 10,000, and the number of *births* 1,000, the *population* increases to 11,000. If the *number of births* then increases to 1,100, the *population* increases further to 12,100, verifying that the link from *births* to *population* is direct, as shown in the diagram. But if the *number of births* falls to 900, the *population* does not fall, but continues to increase, albeit more slowly than it would have done otherwise, to 13,000. This is an example, as mentioned on page 14, of a one-way direct link (technically known as a 'unidirectional inflow' – see page 312); similarly, the link from *deaths* to *population* is a one-way inverse link (a 'unidirectional outflow').

...Depends on the Difference between the Birth and Death Rates...

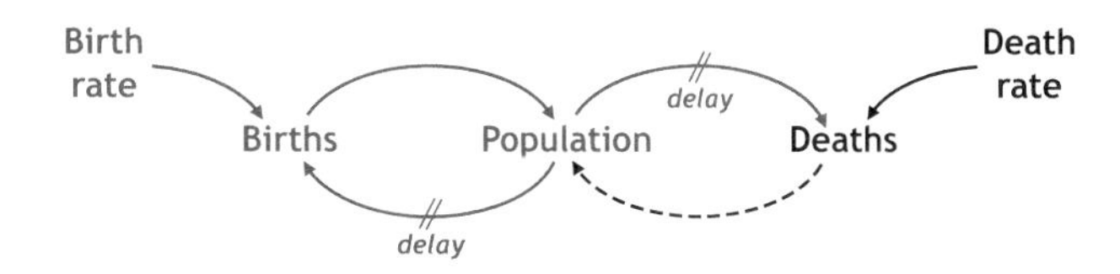

The behaviour of the *population* is determined by the interplay between the reinforcing loop of *births* driving exponential growth, and the balancing loop of *deaths*, holding things back.

The key parameters are the instantaneous values of the *birth rate* and the *death rate*, which together define the *net growth rate* = *birth rate* – *death rate*.

Accordingly, if the *birth rate* is greater than the *death rate*, over any time period, there are more *births* than *deaths*, and the *population* grows exponentially at the *net growth rate* = *birth rate* – *death rate*.

If the *birth rate* is equal to the *death rate*, over any time period, the number of *births* equals the number of *deaths*, and so the *population* remains stable. The *net growth rate* = *birth rate* – *death rate* is zero.

If the *birth rate* is less than the *death rate*, over any time period, there are more *deaths* than *births*, and the *population* declines exponentially at the now negative *net growth rate* = *birth rate* – *death rate*.

In essence, the population system behaves as a single reinforcing loop, as shown to the right. This system can show exponential growth if the *net growth rate* is positive, exponential decline if the *net growth rate* is negative, or remain stable when the *net growth rate* is zero.

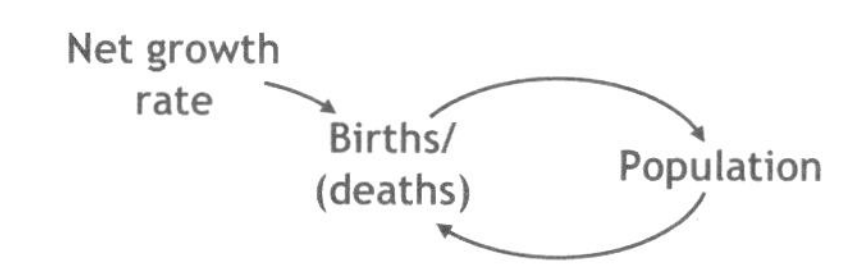

...and Can Appear to Be Very Complex

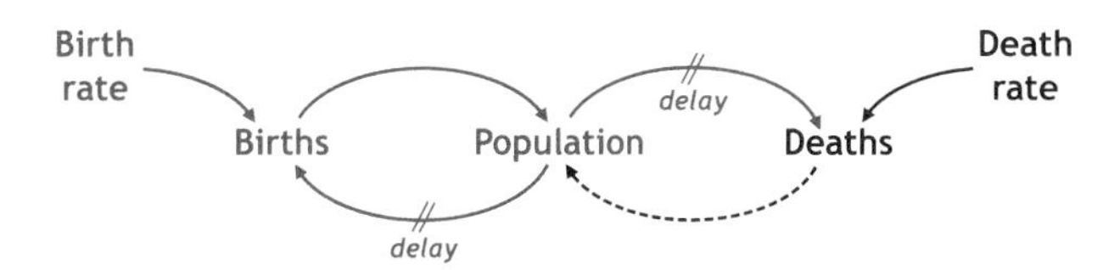

It is usually more informative to express the population system as a reinforcing loop coupled to a balancing loop, for this throws the spotlight onto the factors that influence the *birth rate* (such as, for us humans, the *education of women*) and the *death rate* (for example, the incidence of *disease*). We'll examine this in more detail in Chapter 17 – for the moment, the 'take-home message' is that the actual behaviour of any *population* over time can appear to be very complex, as the following charts illustrate. All are based on simulations of this causal loop diagram, using different instantaneous values of the *birth rate* and the *death rate*.

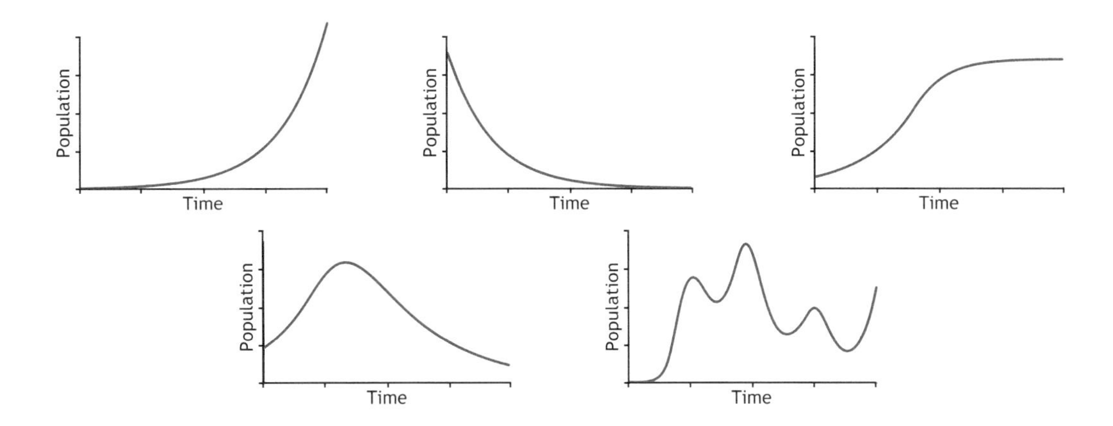

This emphasises how powerful systems thinking can be, showing that many different, and some very complex, behaviours can all be explained by one, simple, causal loop diagram.

Back to the Pond Weed

The nasty-looking green stuff, as a living organism, must be growing exponentially, with, as yet, no constraint.

If the entire surface of the pond is covered by the end of day 20, it is half-covered at the end of day 19, for over the next 24 hours, the number of organisms will double, covering the whole pond. So waiting until the pond is half-covered on day 19 is way too late: the pond will be overwhelmed long before the gardener can apply the remedy – a remedy, which, in systems thinking terms as described on the previous page, acts to increase the death rate of the green stuff.

The latest time at which the gardener can take action by ordering the remedy is on day 10. At which point, the extent of coverage of the pond's surface is at most (1/2) to the power of 10, which is 1/1,024, or less than 0.1% of the surface. The 'signal' that the gardener needs to respond to is therefore very weak – and if the gardener leaves ordering the remedy any longer, the pond will be destroyed, and there is nothing the gardener can do.

A feature of exponential growth is that it starts very slowly, and then … boom!

The world experienced that with Covid-19, and the populations of those governments that waited too long before they locked down ('we need more evidence…') suffered accordingly.

The natural growth rate of Covid-19 was very fast, so the exponential growth was evident quite quickly. Other natural systems that can be described by reinforcing loops might have much slower growth rates, so encouraging 'let's wait and see…'. With the very great danger that remedial action is taken far too late – a story to be told in the last chapter, on the climate crisis…

England's Winter 2020–21 Covid-19 Lockdown...

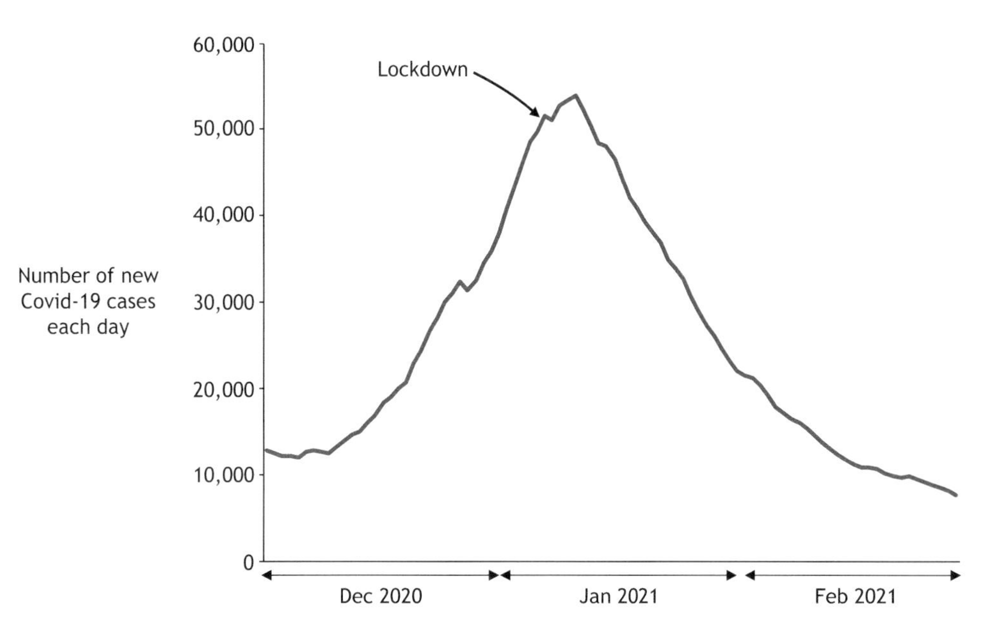

Source: https://coronavirus.data.gov.uk/details/cases.

This chart shows the number of new Covid-19 cases reported in England each day between 1 December 2020 and 28 February 2021, expressed as a 7-day rolling average.

This is very similar to the 'boom and bust' shown on page 43, and also at the lower left on page 58. This shape suggests an underlying reinforcing loop, initially driving exponential growth, and then exponential decline as caused by an external shock, the lockdown declared on 6 January 2021, and which took a few days to show its impact.

Also, on 4 January 2021, the vaccination programme began in England. Since it takes several weeks for the vaccine to become effective, the reduction in cases in January and February is attributable to the lockdown, not the vaccine.

The Covid-19 Reinforcing Loop

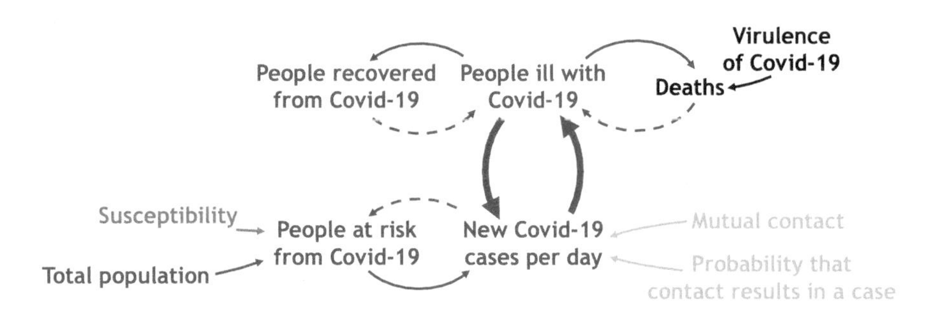

The spread of the Covid-19 virus requires *mutual contact* between a *person who is already ill* and a *person who is at risk*. Not all *contacts* result in infection, and so two other important factors are the *probability that a contact results in a case*, and the *susceptibility to infection* of each individual.

As the central part of the diagram shows, the number of *new Covid-19 cases per day* is both driven by, and a driver of, the number *of people ill with Covid-19*, so forming a reinforcing loop which will exhibit either exponential growth or exponential decline.

There are also three associated balancing loops. First, the number of *new cases* depletes the number of *people at risk*, in that for those unfortunate people, the risk has crystallised. Second, those who *recover* reduce the number of *people who are currently ill*. And third, some of those who have fallen ill tragically *die*, depending on the *virulence of the Covid-19 virus*, and, of course, on the quality and efficacy of medical care. This diagram is therefore a development of the 'figure-of-eight' diagram discussed on pages 56–58.

At the start of the outbreak, in every country around the world, there was much *mutual contact* on public transport, at social gatherings and at work. Furthermore, the nature of the virus was such that there was a high *probability that a contact results in infection*, and a very large proportion of the *total population* was highly *susceptible*. These factors combined to trigger the reinforcing loop, causing the numbers of *new cases* and *people ill* to grow so quickly that many individuals suffered greatly, and healthcare services soon became over-stretched.

Lockdown, Face Masks, Hand-Washing and Vaccine

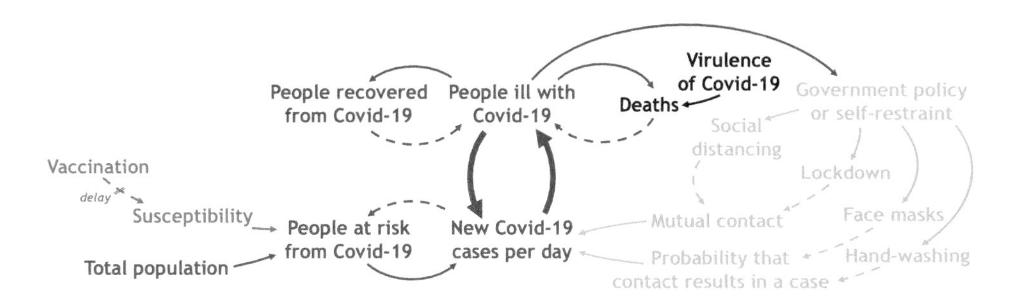

The *probability that contact results in a case* can be reduced by wearing a *face mask* and by thorough *hand-washing*, and *mutual contact* can be reduced by *social distancing*, and – more stringently – by a community *lockdown*. These can each be implemented by individuals *voluntarily*: some people might choose, for example, not to visit a sporting event, imposing their own personal *lockdown*, perhaps as the result of the fear of infection attributable to the rapidly increasing number of *people already ill*. Also, as the strain on healthcare services intensified, *governments* actively publicised the importance of *social distancing*, wearing *face masks* and *hand-washing*, as well as enforcing *lockdowns*, perhaps reluctantly, and often against opposition.

Structurally, these actions introduce four further balancing loops, each of which acts to slow down the exponential growth of the central reinforcing loop. And collectively, they caused the action of the loop to flip from exponential growth to exponential decline, as illustrated by the actual data for England over the winter months of 2020–21, as shown on page 60, and also in the simulation results presented on page 63.

Vaccination reduces *susceptibility*, so lowering the *risk of infection* – but the development and testing of a new vaccine takes a *considerable time*, and even when a safe and effective vaccine is available, it takes *more time* for whole populations to receive the required number of doses.

A Simulation of the Real Data

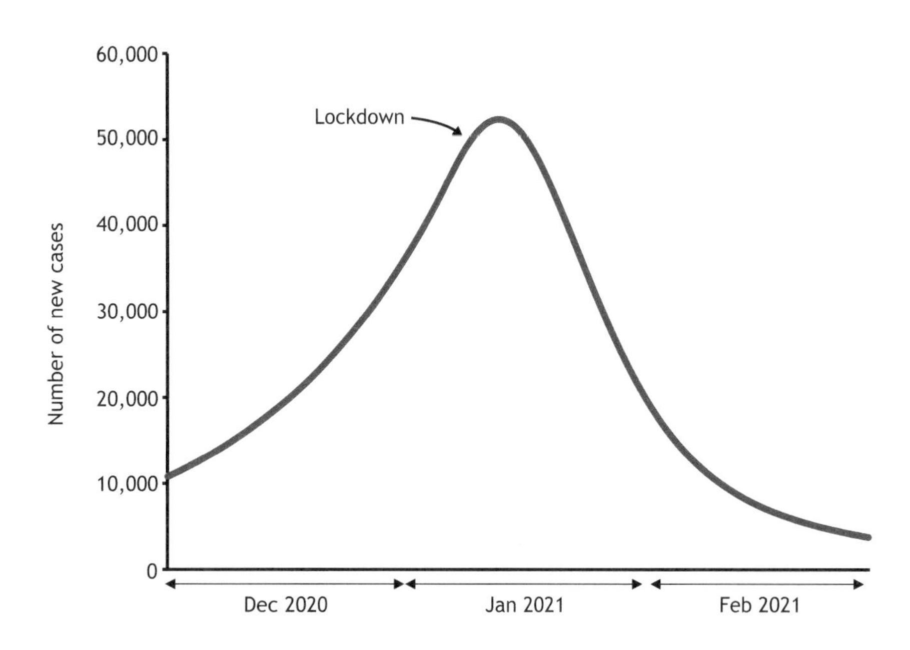

This is the author's computer simulation of the causal loop diagram shown on page 62, before the *vaccine* takes effect. The resulting shape is very similar to the actual data as shown on page 60.

This causal loop diagram is very simple, and ignores many real factors. None the less, the simulation of the actual shape is good, demonstrating that even simple causal loop diagrams can give considerable insight into real, complex, systems.

Mutations and Immunity

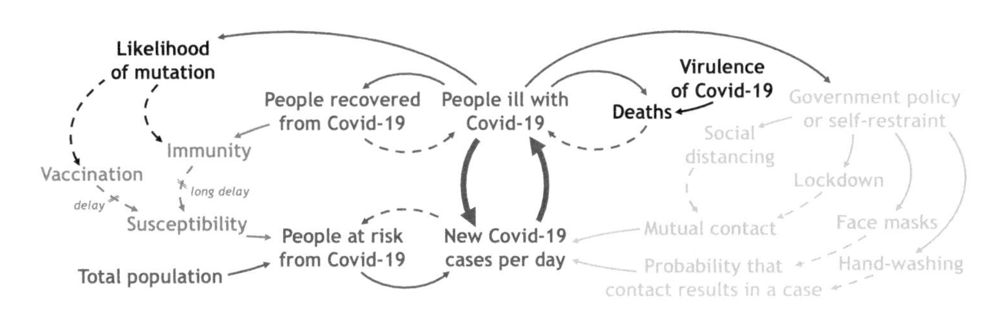

Some of those who *recover from Covid-19* have some natural *immunity*, which acts to reduce their *susceptibility*, so putting them at much less *risk*. In principle, this introduces a balancing loop which acts to slow down the reinforcing loop of *new cases* and *illness*. However, for '*herd immunity*' to have a noticeable effect takes a long time, and requires that a large proportion of the *total population* has suffered the disease, with the corresponding distress and number of *deaths*.

Also, the greater the number of *people ill with Covid-19*, the greater the *likelihood of mutation* of the virus. Some of these *mutations* might be resistant to the existing *vaccines*, so reducing their effectiveness, as well as possibly diminishing the *immunity* of those who had previously recovered. This introduces two further reinforcing loops, which – in the absence of a *lockdown* or the rapid development of a new *vaccine* – act to increase the number of *new cases*, as actually happened with the emergence of the 'delta' and 'omicron' mutations that drove the subsequent 'waves' of the pandemic during 2021 and 2022.

A full analysis of Covid-19 requires a much more complex causal loop diagram, with many more variables. This quite simple diagram, however, captures the essence of the pandemic, and consists of a network of three reinforcing loops and eight balancing loops, all of which are operating simultaneously to their own appropriate rhythms. Even such a relatively simple system can exhibit complex behaviour – but behaviour that can be understood, and complexity that can be tamed.

Chapter 6

Targets and Budgets

DOI: 10.4324/9781003304050-7

Alex's Big Opportunity

'The business is really growing well, and next year looks really strong.'

'That's great – we've all worked very hard.'

'Yes, we have. And there's even more to be done next year, so it makes sense for your team to expand. I think you should increase its size from the 12 you have now to about 15 in 3 months' time. Does that make sense to you?'

'Sure does. I'll liaise with HR right away to get some recruitment going. Thank you!'

Closing the Headcount Gap

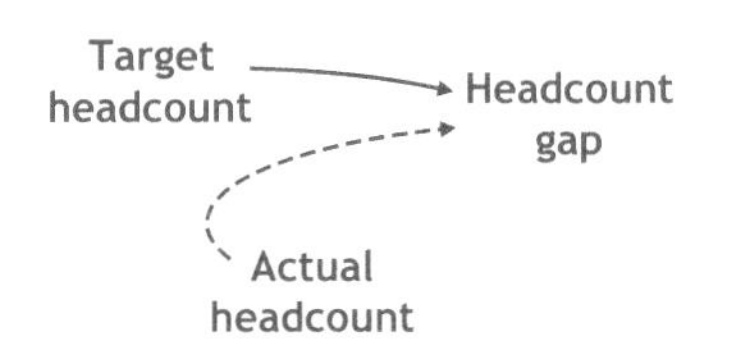

The *actual headcount* in Alex's team is current 12, and the boss has just authorised a new *target headcount* of 15. This opens a *headcount gap*, this being the difference between the *target headcount* and the *actual headcount*:

Headcount gap = Target headcount – Actual headcount

These relationships can be represented by the causal loop diagram to the left, in which the link from *target headcount* to *headcount gap* is direct, and the link from *actual headcount* to *headcount gap* is inverse.

The *headcount gap* is a trigger for *recruitment*, and the larger the *gap*, the more extensive the *recruitment* campaign, so that's a direct link too, as shown to the right.

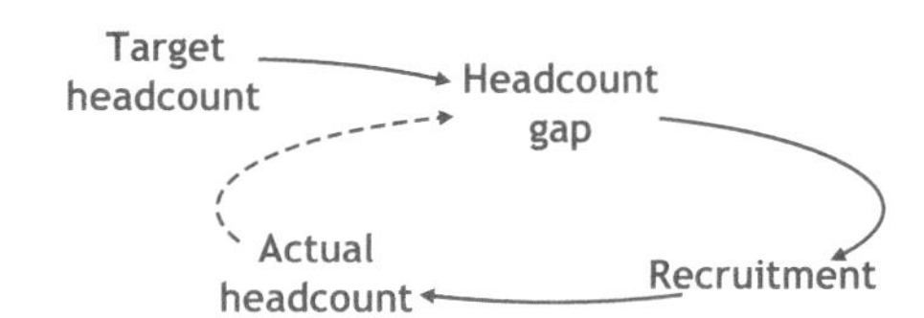

Every new person adds to the *actual headcount*…

…so progressively reducing the *headcount gap*…

…until, when the *actual headcount* equals the *target headcount*, the *headcount gap* becomes zero, and so the *recruitment* becomes zero too, and the system stops.

A Different Form of Balancing Loop…

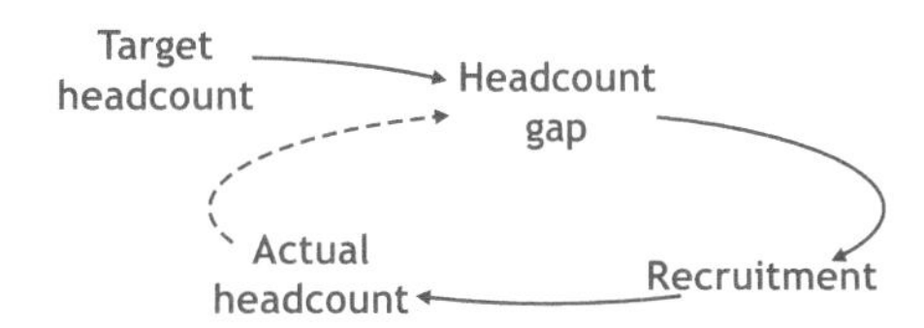

This causal loop diagram describes a system that is designed to achieve a *target*, to reach a goal: setting a *target* opens a *gap* with the corresponding *actual*, triggering an appropriate *action* to do whatever is required to bring the *actual* into line with the *target*, at which point the system 'switches itself off'.

The closed feedback loop has one inverse link, and since one is an odd number, this loop is a balancing loop.

But unlike the balancing loops discussed on pages 53, 55 and 56, this balancing loop is not coupled to an associated reinforcing loop, but rather is associated with a **target dangle**.

...Which Can Act 'Both Ways'

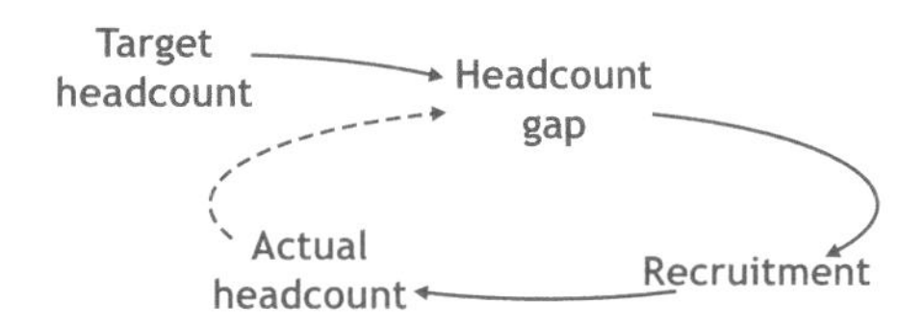

Suppose that the *actual headcount* has risen from 12 to 15, in accordance with the *target headcount*. But suppose further that as a result of a business downturn, Alex is told that the team is now too large, and to save costs, must be reduced to 10.

The new *target headcount* is now 10, and so the *headcount gap* = *target headcount* – *actual headcount* = 10 – 15 = –5 is now a negative number. In accordance with the direct link from *headcount gap* to *recruitment*, this then drives the 'negative *recruitment*' of –5 people, which, in more every-day language implies that five people are made redundant. This 'negative *recruitment*' then results in a reduction in the *actual headcount* to ten people. At this point, the *actual headcount* is once more in line with the *target headcount*, the *headcount gap* reduces to zero, and the system once again 'switches off'. The causal loop diagram therefore can work 'both ways', catering for both hiring and firing, if the term *recruitment* can be interpreted broadly, and allowed to be associated with both positive and negative numbers; alternatively, the diagram can be represented as shown below:

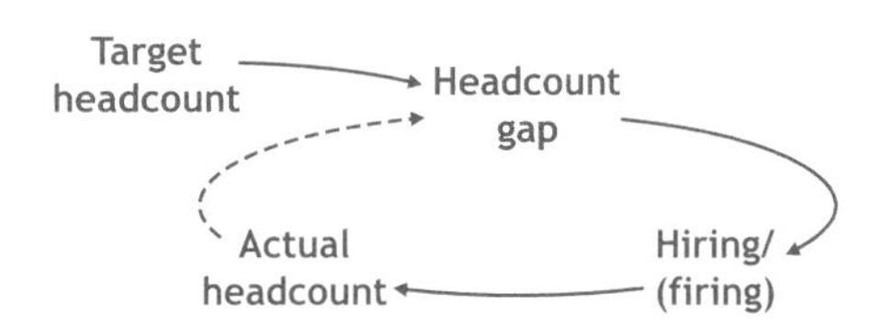

Targets, Budgets and Performance Measures

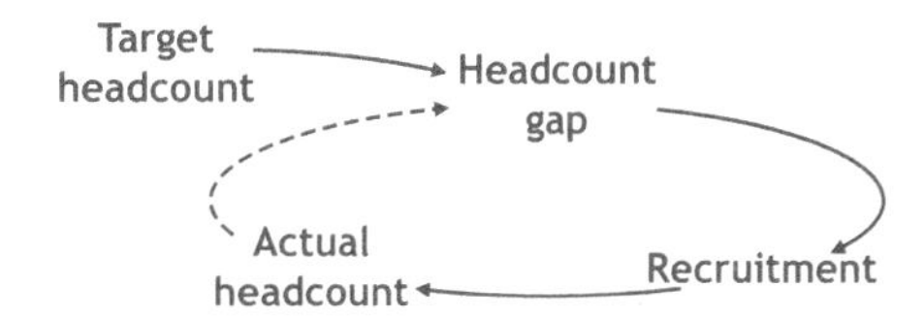

The example of the departmental *headcount* is a particular instance of the much more general context of targets, budgets, objectives, goals, performance measures, ambitions and aspirations. Wherever these exist, whenever there is something I want to do, am being encouraged to do, or am obliged to do, there is some form of *target*, a goal to be achieved, an aspiration that is desired. Inevitably, the *target* is different from the current *actual* state, so opening up some form of *gap* which then triggers whatever *action* is required to bring the *actual* into line with the *target*. When this is achieved, the *gap* reduces to zero, implying that no further *action* is required. Accordingly, the *actual* remains stable, and the same as the *target* for as long as the *target* itself remains stable, as illustrated in the graph to the left, below.

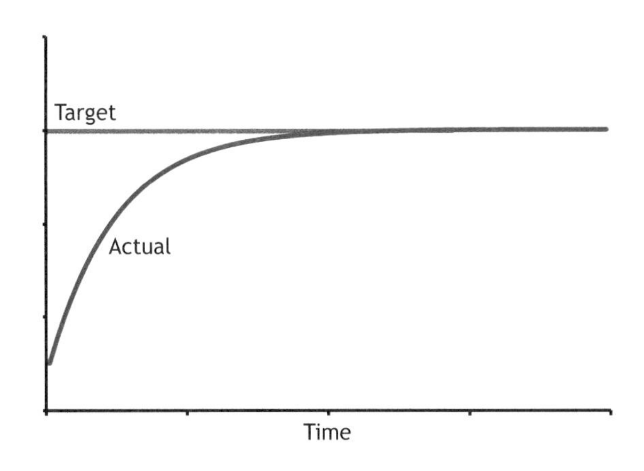

In this example, the *target* is expressed as a target dangle, representing a *target* introduced into the system from 'the outside'. This is often the case, and many causal loop diagrams include target dangles. In reality, of course, *targets* themselves can be determined from other factors, and so it is quite possible for the *target* associated with any particular balancing loop to be connected to other variables, and so not a dangle. None the less, such a *target* will still define a *gap* with respect to a corresponding *actual*.

The Problem of Delays

All balancing loops associated with a *target* act to bring the *actual* into line with the *target,* and for systems that react quickly, the *target* is approached smoothly, as shown in the chart on the previous page.

In some systems, however, there can be *delays* – for example, *delays* in measuring what the *actual* is (implying that the *gap* is identified some time later, by which time the value of the *actual* might have changed), or *delays* in taking *action* (such as dithering over whether any *action* should or should not be taken), or *delays* associated with the time taken before the results of the *action* taken take effect.

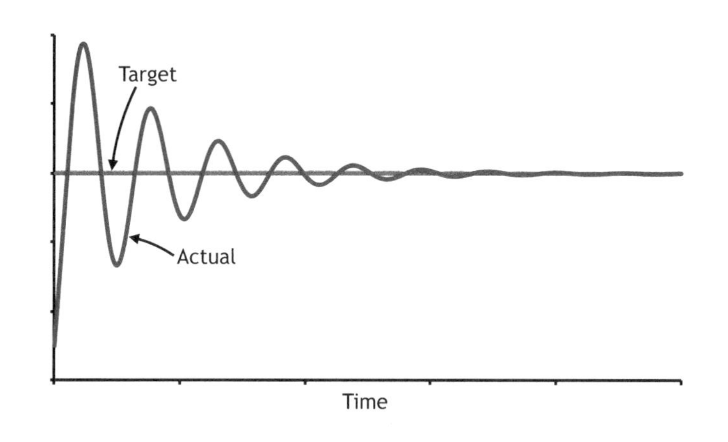

The effect of a delay is exemplified by the chart to the left: the system over- and under-shoots until it stabilises.

An example is what happens when a car, driven by a relatively inexperienced driver, is in a skid: by the time the driver reacts, the car has already moved to a different position, and so the setting of the steering wheel is wrong. The driver then over-reacts in the other direction, and the car lurches from side to side until the driver regains control.

This serves as a warning to any manager trying to control a system subject to delays: the manager might react to the initial overshoot, for example, by changing the target – which might make matters even worse. The message here is 'be patient', or – even better – redesign the system to remove the delays.

The Definition of 'Gap' Is a Convention…

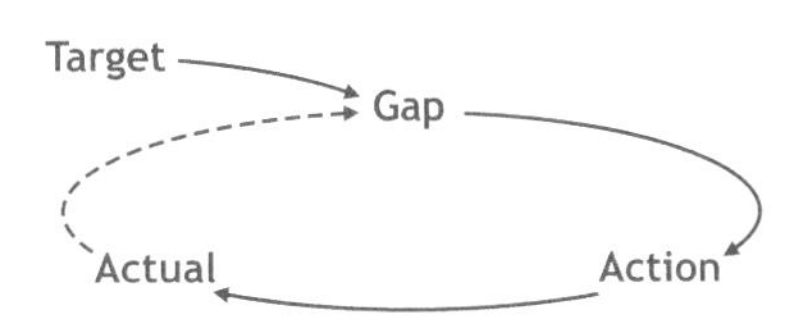

The balancing loop examined so far, as shown to the left, defines the *gap* as

$$Gap = Target - Actual$$

Accordingly, the *gap* is positive when the *target* is greater than the *actual*, and negative when the *target* is less than the *actual*. This is the formal definition of what accountants call 'variance', and is the 'natural' way to express the *gap* when 'big is good', and *actuals* are aspiring to meet higher *targets*.

This definition, however, is just a convention – the *target* can also be defined the other way around as

$$Gap = Actual - Target$$

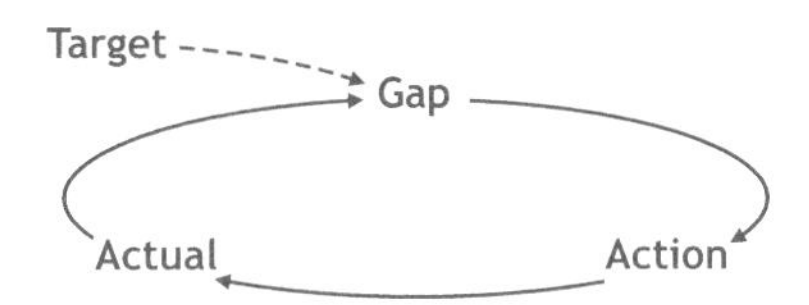

as is often appropriate when 'small is good' – for example, if the *target* is a delivery or service time, such as the time someone phoning an enquiry centre has to wait until the call is answered.

A consequence of this alternative definition is that the link from *target* to *gap* is now inverse, and the link from *actual* to *gap* is now direct, as shown to the right.

This diagram, however, is problematic, for the closed feedback loop from *gap* to *action*, to *actual*, and back to *gap*, is now composed of three direct links and no inverse links, implying that this is now a reinforcing loop…

…which just can't be right – a system cannot change its behaviour from acting as a balancing loop to acting as a reinforcing loop simply because one particular variable, the *gap*, is defined the other way around. So that diagram must be wrong.

...Two Other Forms of Balancing Loop

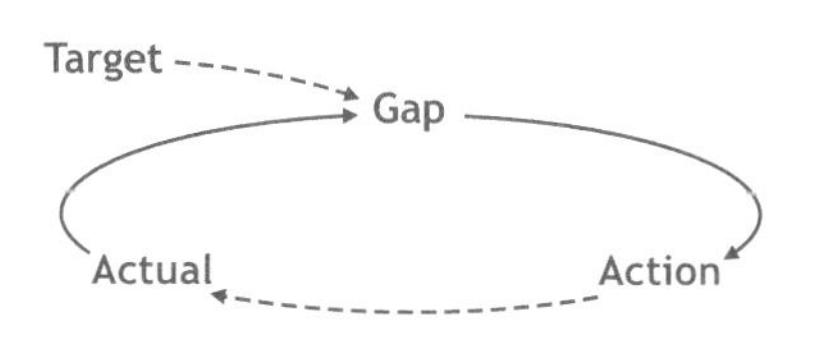

The problem identified on the previous page can be resolved by recognising that when the link from *actual* to *gap* changed from inverse to direct, the link from *action* to *actual* must change from direct to inverse. This results in a closed loop containing one inverse link, which is a balancing loop, as required.

This is best illustrated by example, so suppose that a medical laboratory has a *target* that the results of all diagnostic tests must be despatched within 48 hours of receipt of the test samples. Suppose further that the laboratory has been short-staffed, causing a backlog to build up with the result that the despatch time *actually achieved* has risen to 54 hours, implying that the *gap* = *actual* – *target* = 54 – 48 = 6 hours.

This triggers some appropriate *action,* for example, overtime working, to reduce the backlog, bringing the *actual* back down to 48 hours. An increase in *action* has resulted in a decrease in the *actual*, and so the corresponding link is inverse.

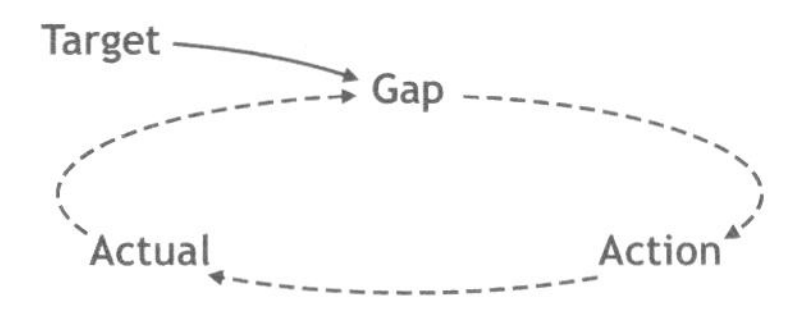

Suppose, however, that the original definition of the *gap* is maintained as *gap* = *target* – *actual*, implying that the link from *target* to *gap* is direct, and the link from *actual* to *gap* is inverse.

If the *actual* despatch time is 54 hours and the *target* is 48 hours, then the *gap* = *target* – *actual* = 48 – 54 = –6 hours, a negative number.

This negative *gap* then drives the positive *action* of overtime, which, in turn, reduces the *actual* 48 hours.

The links from *gap* to *action*, and from *action* to *actual*, are therefore both inverse. The closed feedback loop is composed of three inverse links, and since three is an odd number, the loop is a balancing loop.

Balancing loops may therefore take a variety of forms, depending on the definition of *gap*, and whether 'small' or 'big' is 'good'. But all exhibit the fundamental feature of containing an odd number of inverse links.

A Special Case

A balancing loop that is not associated with either a coupled reinforcing loop, or with a *target*, and so is just 'there, by itself' is, in my experience, extremely rare. The only instance I know concerns the world of physics, and the to-and-fro motion of, for example, a pendulum, which physicists and mathematicians refer to as 'simple harmonic motion'...

...in which, at any time, three key attributes of the system can be measured: first, the *displacement* of the pendulum with respect to the central position (at which the pendulum would be at rest, if it were not swinging to and fro); second, the corresponding *acceleration*; and thirdly, the *force* exerted on the pendulum by gravity.

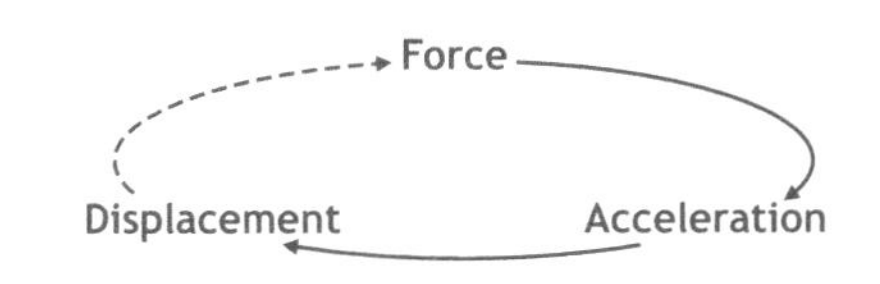

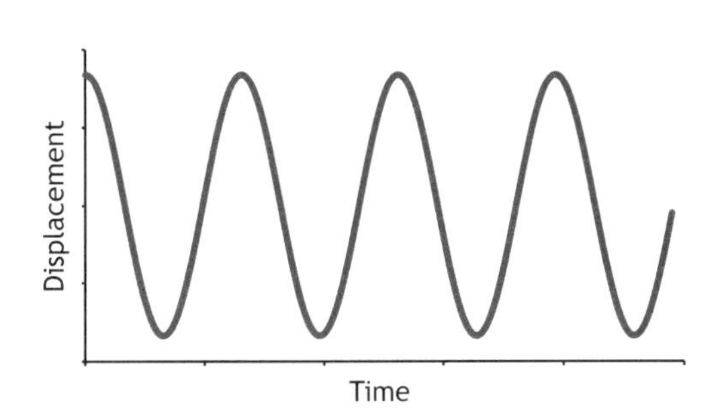

According to the laws of physics (which I won't go into here!), the greater the *force*, the greater the *acceleration*, so that's a direct link. The greater the *acceleration*, the farther the pendulum will travel over any given time, so the greater the *displacement*, implying another direct link. But the greater the *displacement*, the stronger the *force in the opposite direction*, opposing the motion of the pendulum. Mathematically, that is represented as a negative number, hence the inverse link.

The resulting system is a 'pure' balancing loop, with no associated *target*, and no associated coupled reinforcing loop.

And its dynamic behaviour is to oscillate forever...

Balancing Loops – Summary

A balancing loop is any closed feedback loop in which the number of inverse links is odd, and it shows one of three types of dynamic behaviour...

If the balancing loop is associated with a coupled reinforcing loop, the action is to slow the rate at which the reinforcing loop exponentially grows or declines, in effect controlling the reinforcing loop.

If the balancing loop is associated with a target – which is often, but not necessarily, in the form of a target dangle – the action is to bring the corresponding actual into line with the target, either directly and smoothly, or – if there are time delays in the operation of the feedback loop – after a series of over- and under-shoots. A balancing loop of this type is associated with all targets, objectives, goals, performance measures, budgets, ambitions and aspirations, and so is encountered very frequently.

A 'stand-alone' balancing loop – one which is not associated with either a target or a coupled reinforcing loop – is very rare, and its behaviour is an infinite oscillation.

The 'Fundamental Theorem' of Systems Thinking

Every individual link in any closed feedback loop must be either a direct link or an inverse link.*

Any single closed feedback loop must contain either an even, or odd, number of inverse links.

Therefore every closed feedback loop must be either a reinforcing loop or a balancing loop.

* As noted on page 30, influence links can ONLY be used for dangles and must NEVER be used within a closed feedback loop. The reason for this is now clear, for this 'fundamental theorem' relies on this prohibition.

Why the 'Fundamental Theorem' Is Important

Real systems are complex.

We all know that. That's why they're hard to predict, hard to understand and hard to manage.

All real systems, however, can be represented by well-thought-through causal loop diagrams. But because the systems being described are complex, these diagrams should be expected to be complex too – as indeed will be verified by the examples in later chapters of this book.

All causal loop diagrams, however, are networks of interconnected closed feedback loops, with a (usually quite small) number of associated dangles.

The 'fundamental theorem' tells us that each of those closed feedback loops is either a reinforcing loop or a balancing loop. **There Are No Other Possibilities**.

Furthermore, reinforcing loops and balancing loops are easy to identify, and to distinguish. Just count the number of inverse links with the closed loop. The result can be any number, a small one or a big one. But whatever that number might be, that number must be either even or odd. **There Are No Other Possibilities**.

And if the number is even, the corresponding closed loop is a reinforcing loop; if odd, a balancing loop.

This makes it much easier to understand how real systems behave.

Yes, the system may be (and probably is) hugely complex, but that complexity is an assembly of just two, very fundamental, 'building blocks' – reinforcing loops and balancing loops. And if you have a good appreciation of how each of these behaves, then understanding how they all fit, and operate, together in a real system becomes a much more feasible proposition. Especially when the fundamental behaviour of each type of loop is itself simple – reinforcing loops either grow or decline, balancing loops either slow down an associated coupled reinforcing loop, or act to bring some form of *actual* into line with the corresponding *target*.

A Second Time to Pause...

From here, the book changes up another gear, combining the key building blocks of causal loop diagrams – **reinforcing loops** and **balancing loops** – to describe a wide variety of real systems. So the key pages are 48 (for reinforcing loops), 75 (for balancing loops), in the context of the **fundamental theorem** (page 76).

So when you're happy with all that, turn the next page and hold your hat!

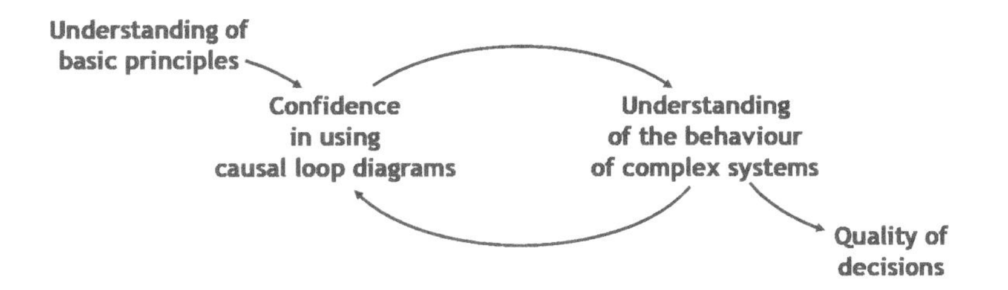

PART 2 Applications

Competitive Markets

'I've been doing some research, and I think our margins are the best in the business. That means we can reduce our prices to steal some market share, and the competition won't be able to retaliate because their margins would then be unsustainable. The rational decision for them must be to accept the loss of some market share for a while, and work on improving their margins so they can match our prices sometime in the future. But in the meantime, we will have made a killing!'

'Maybe. But is their only way to retaliate by matching our price?'

Chapter 7

Competitive Markets

DOI: 10.4324/9781003304050-9

Red's Business...

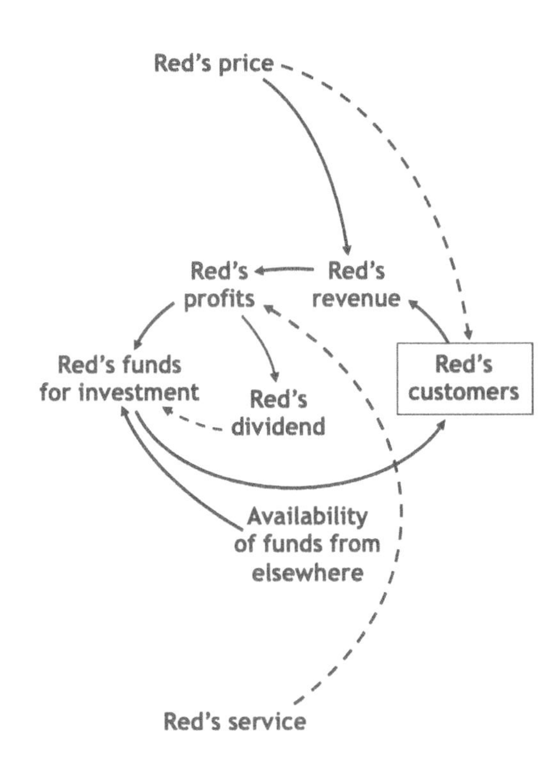

Red runs a *service* business that can be described by the now-familiar reinforcing loop: *Red's customers* generate *revenue* and *profits* which provide the *funds available for investment* that makes the business even more attractive, so increasing the *customer base.*

The *profits* also provide the *dividend*, which takes priority over the *investments.* Accordingly, the *funds for investment* are determined by what's left after the *dividend* has been paid, and so

Funds for investment = Profits – Dividend

That explains the inverse link from *Red's dividend* to *Red's funds for investment*; had the priorities been the other way around, implying that

Dividend = Profits – Funds for investment

then the inverse link would have been from *Red's funds for investment* to *Red's dividend.*

In general, *funds for investment* can be obtained from sources other than *customers*, for example from shareholders or loans, as indicated by the *availability of funds from elsewhere.*

As discussed on page 31, an increase in *Red's price* increases *revenue* by virtue of the *revenue* = *price* × *volume* relationship; at the same time, a price increases acts as a disincentive to *customers* – hence the inverse link from *Red's price* to *Red's customers.*

Finally in relation to this diagram, the costs of delivering *Red's service* are represented by the inverse link to *Red's profits.*

...and Blue's

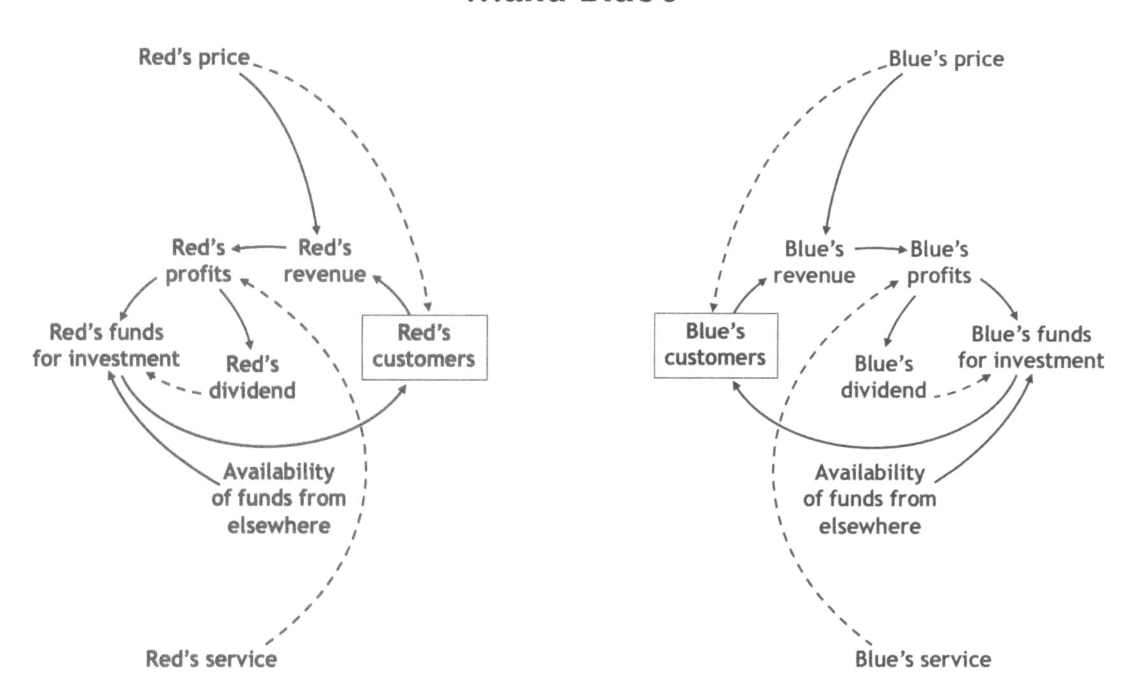

Blue runs a similar business, which, as shown on this causal loop diagram, does not compete *with* Red...

Red and Blue Compete on Price…

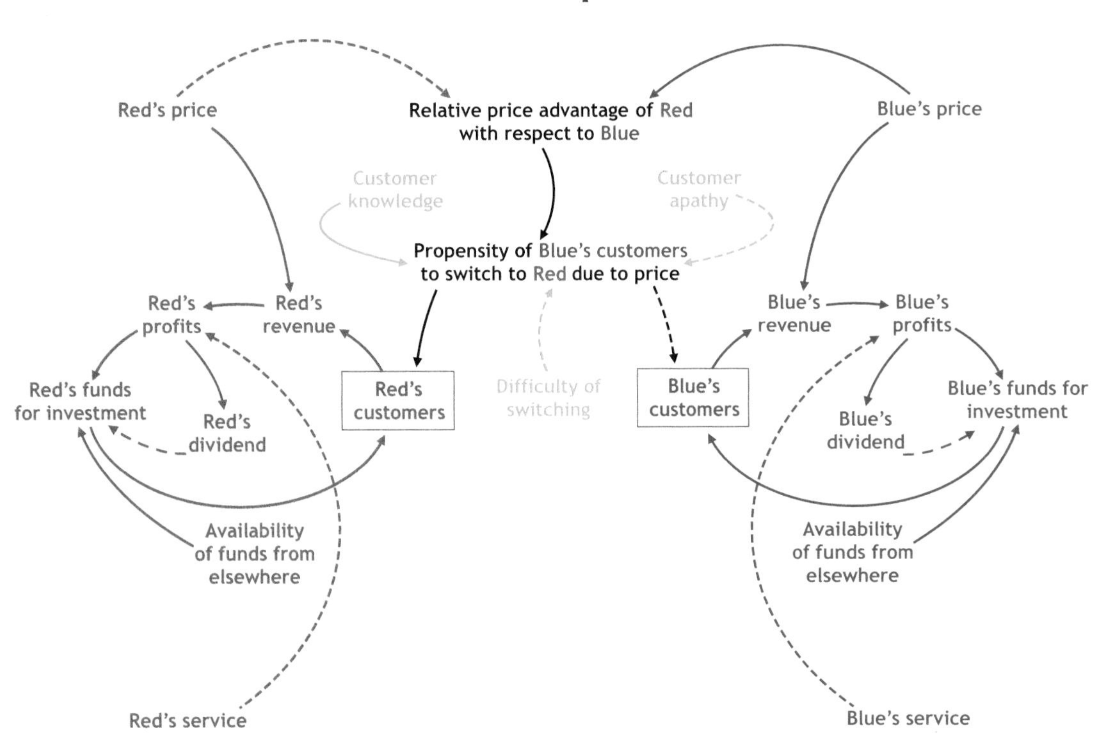

For narrative, see following page

Red and Blue Compete on Price...

But if *Red* and *Blue* do compete, then perhaps *price* might be an important factor in influencing *customer choice.*

If the *relative price advantage of Red with respect to Blue* is defined as *Blue's price – Red's price,* then this is a positive number when *Red's price* is lower than *Blue's price,* giving *Red* an advantage...

...likewise, if it is a negative number, *Red's price* is higher than *Blue's price,* giving *Blue* the advantage.

A *relative price advantage in favour of Red* is likely to increase the *propensity of Blue's customers to switch to Red due to price,* so increasing *Red's customers* whilst simultaneously depleting *Blue's customers* – that's why there is a direct link from *propensity* to *Red's customers* and an inverse link from *propensity* to *Blue's customers.*

Similarly, a *relative price advantage in favour of Blue* (as represented by a negative number) causes the *propensity of Blue's customers to switch to Red,* to become negative. Since the link to *Red's customers* is direct, this causes a decrease in *Red's customers,* and the inverse link to *Blue's customers* causes this to increase. This is a switch of customers from *Red* to *Blue,* as would be expected as a consequence of a *relative price advantage in favour of Blue.*

This causal loop diagram therefore represents how the pricing policies of *Red* and *Blue* can influence *customer behaviour,* and result in *switching,* but only if the *customer knows* about the *price differential* as well as being sufficiently *concerned* to take the effort to *switch,* and to navigate any *procedural difficulties* that might be in the way – as can be the case, for example, for utilities such as gas, electricity and mobile telephones.

The sequence of three links from *Red's price* to *relative price advantage, propensity to switch* and then *Red's customers* is inverse, direct, direct, which collectively behave as a single inverse link. Together, these three links replace the single inverse link from *Red's price* to *Red's customers* as shown on pages 82 and 83, and preserve the general principle that an increase in *price* is likely to result in a decrease in *customers.*

This applies to *Blue* too, for the sequence direct, direct, inverse from *Blue's price* to *relative price advantage, propensity to switch* and then *Blue's customers* also collectively behave as a single inverse link.

If a *customer* considers *Red's price* to be too high and is not attracted to *Blue's* offer, that *customer* might choose not to make a purchase at all. The option not to buy is, in essence, a competitor to *Red,* a competitor with a built-in perpetual *price advantage,* a competitor that is always there, and a competitor to which that customer might choose to switch at any time – provided, of course, that the *customer* can do without the product or service in question.

...and on Service

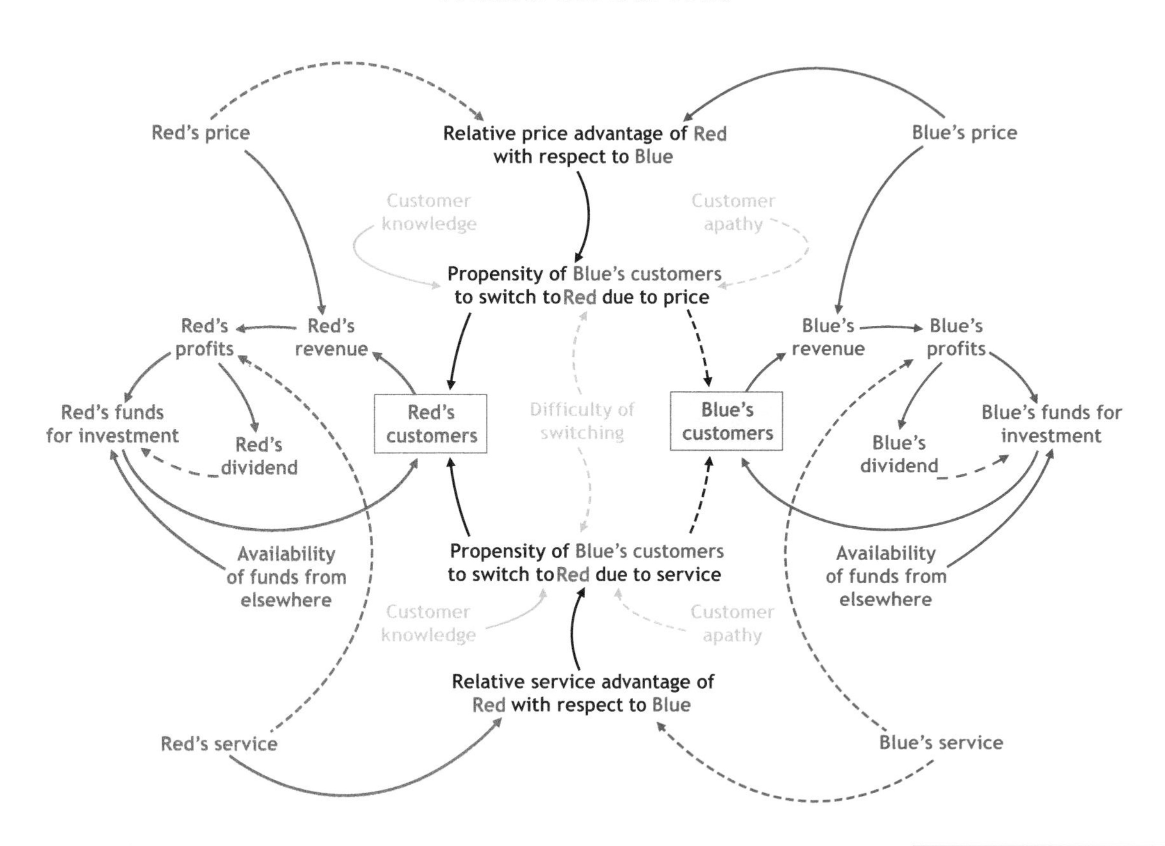

For narrative, see following page

…and on Service

Price is not the only basis of competition between *Red* and *Blue* – another might be *customer service*, for example, delivery time or some form of after-sales support.

Accordingly, there might be a *relative service advantage of Red with respect to Blue*, driving a corresponding *propensity of Blue's customers to switch to Red due to service*, once again increasing *Red's customers* whilst simultaneously depleting *Blue's customers*.

This causal loop diagram is symmetrical for *Red* and *Blue* – which is a pictorial representation of the often real fact that the two competitors are in many ways very similar. Indeed one of the key strategic challenges for any management team is to identify, and maintain, features of the business offer that are valuable to *customers*, and that are not only different from what the competition is offering, but also hard to copy.

If *prices* and *service levels* stay stable over time, the system remains stable too, with both *Red* and *Blue* maintaining their respective *customers*, and with the financial surpluses being held in reserves, spent on activities not directly related to *customers* (such as that new Head Office), or perhaps returned to shareholders as *dividends*.

Red Strikes…

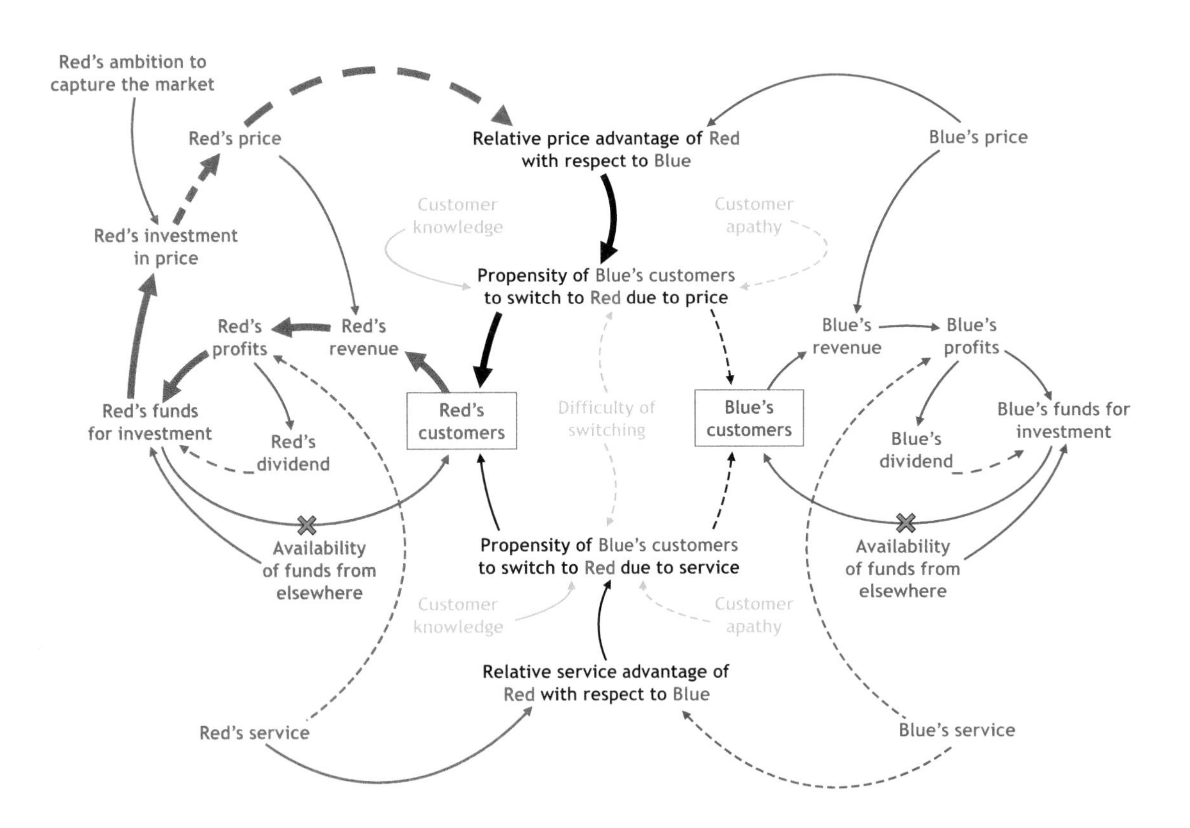

For narrative, see following page

Red Strikes...

But suppose that, in a bid to *capture the market*, *Red's* management decides to lower *Red's price*, with the intention of increasing the *relative price advantage of Red with respect to Blue*, so encouraging *customers to switch*.

By reducing the *price*, *Red's* management is, in the shorter term, deliberately reducing future *profits*, in the hope that this will later be more-than-compensated by an increase in *customers*. This decision is therefore an *investment in price*, the corresponding *funds* being the future *profits* forgone.

And it is an *investment*, for there is no guarantee that *customers will indeed switch* – as the causal loop diagram shows, *customers need to know* that *Red's price* has been lowered, and *apathy* – or sheer inertia – can be a powerful force stopping the *customer* from taking action, even if the action is in the *customer's* interests. So *Red's* management are likely to need to do a number of other things, in addition to just lowering the *price*.

The sequence *funds for investment*, *Red's investment in price*, *Red's price*, *relative price advantage of Red*, *propensity of Blue's customers to switch to Red*, to *Red's customers* is one possible reality of what so far has been represented by the single link from *funds for investment* to *Red's customers*. This single link is therefore superseded by the sequence through *Red's price*, hence the X indicating that this link is no longer operative.

This sequence also forms a major part of a reinforcing loop, triggered by a reduction in *Red's price* in the expectation of causing an increase in *Red's customers*. This will result in exponential growth for *Red* – but at the expense of *Blue*...

...and Blue Retaliates...

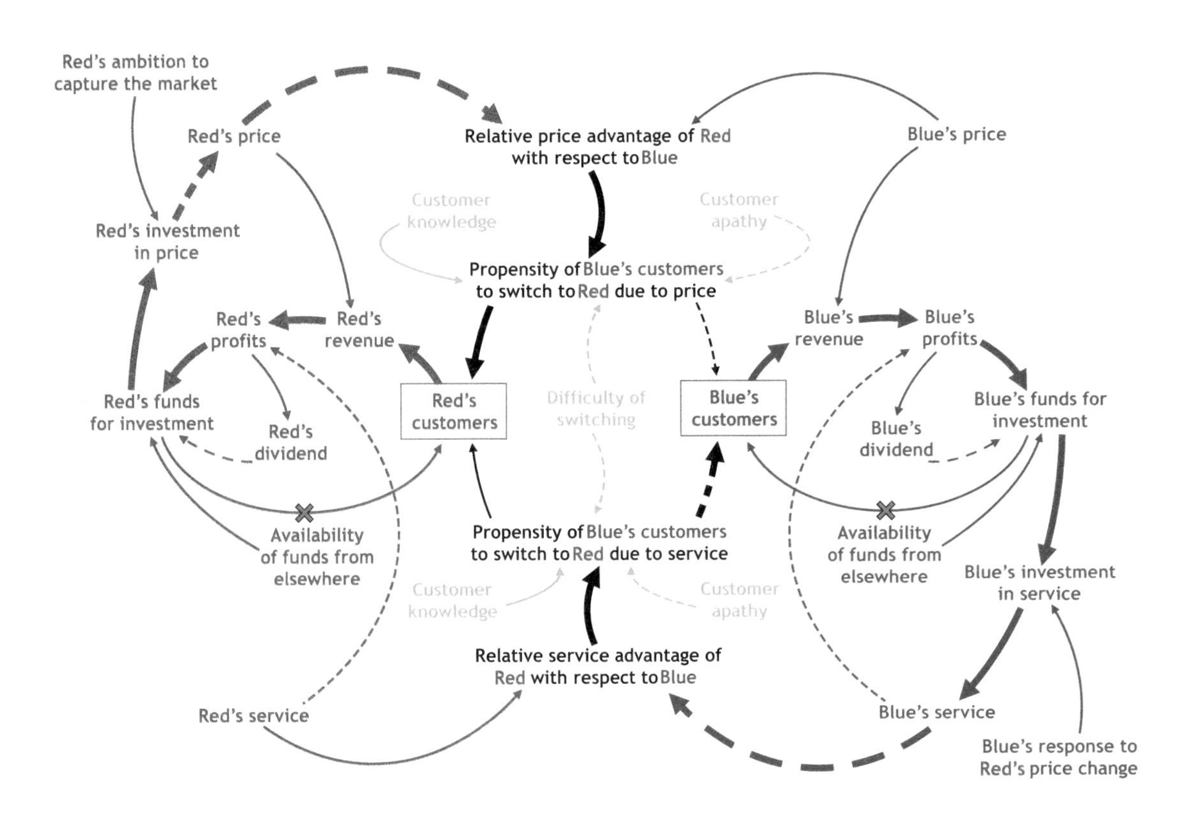

For narrative, see following page

...and Blue Retaliates...

Blue's management will soon notice the loss of their *customers*, and will take action accordingly.

But rather than choosing to drop *Blue's price* to match *Red's price reduction*, *Blue's* management chooses to improve their *service* proposition, and by *investing in service*, *Blue's* management believes that that those *customers* who might have been tempted to *switch* as a result of the *price differential* will choose to remain loyal by virtue of the improved *service*.

Even better, suppose that the *Blue's relative service advantage over Red* is so compelling that *Red's customers switch to Blue...*

The overall system now comprises two, competing, reinforcing loops, one driven by the *relative price advantage of Red with respect to Blue*, the other, by the *relative service advantage of Blue with respect to Red*.

It is possible for such a system to maintain a precarious equilibrium – but such an equilibrium can be unstable. If the total number of *customers* is constrained – as happens when a market is genuinely saturated – it is not possible for both *Red* and *Blue* to grow simultaneously. So a system in which the two loops both show exponential growth won't happen.

Mutual exponential decline is a possibility, but is conditional on either no-one wanting the offer from either *Red* or *Blue* (as might happen if a new technology emerges which neither *Red* nor *Blue* adopt), or if a new competitor, Green, enters the market with a superior offer.

If the system does not maintain equilibrium, the most likely other outcome is that one reinforcing loop shows exponential growth and the other exponential decline, so that either *Red* or *Blue* is driven out of business, or becomes sufficiently small as to not to bother the 'winner'.

This is not hypothetical, as those driven out of business by the internet giants, or by the major retailers, or – in the past – by the 'robber barons' of steel and railways, know all too well.

Who Wins? And Who Loses?

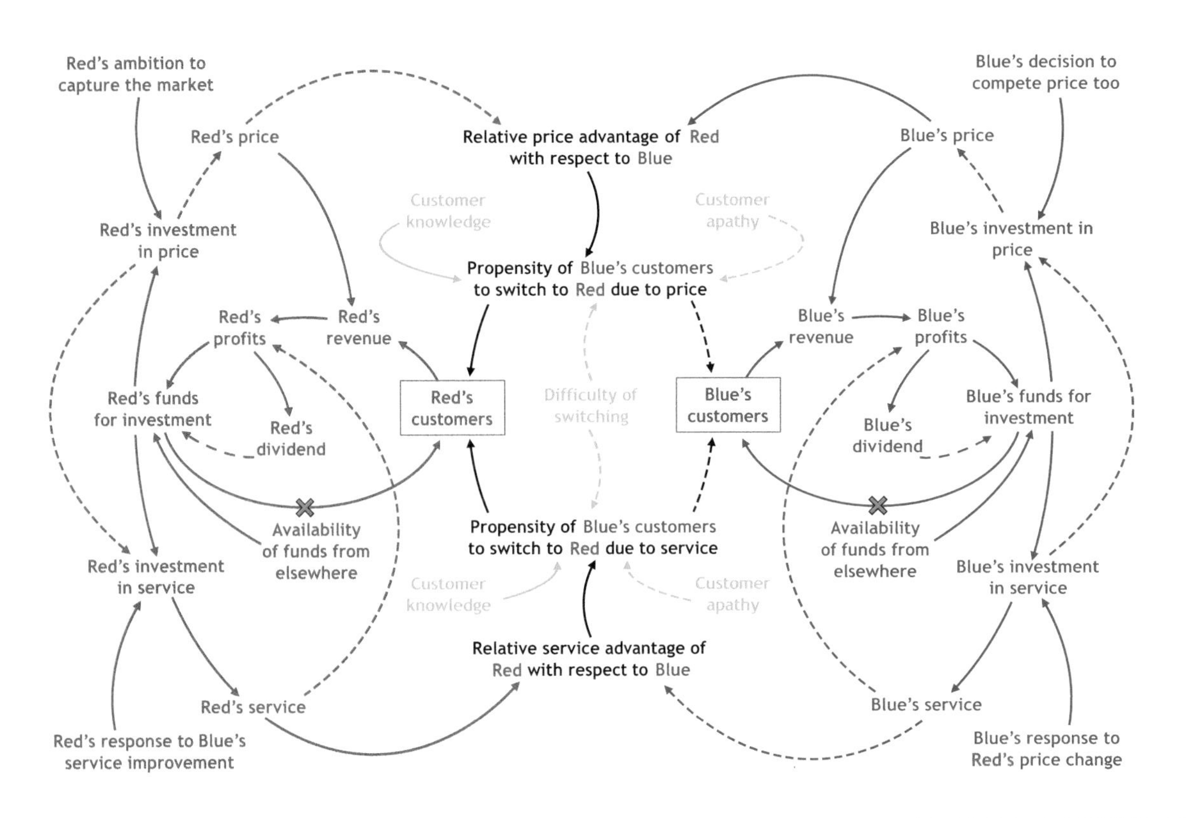

For narrative, see following page

Who Wins? And Who Loses?

But the story does not stop there. As soon as *Red's* management are aware of the *service advantage of Blue*, they attempt to match it, provoking *Blue's* management to retaliate by dropping *Blue's prices* too.

This can be represented by a causal loop diagram shown on page 92, which comprises four interconnected reinforcing loops, all battling to grow exponentially, and to avoid exponential decline.

Matters don't stop there, for *price* and *service* are just two of the many possible bases for competition. Indeed, one of the most important strategic questions is 'what features of our business provide the best opportunities for us to gain a sufficient *relative advantage* for long enough for that exponential growth to make a real difference?'

Perhaps the answer is not 'a feature of our business', but rather a new 'feature' than isn't part of the business yet, but which offers great opportunities; perhaps the answer is to seek new *customers* that are not existing *customers* of a competitor, and who therefore need to be persuaded to *switch*, but are perhaps *customers* of a much weaker competitor, or – better – new *customers* who are the first into a new market…

And, as the causal loop diagram on page 92 illustrates, beware *price* wars. If *Red's price* is reduced, and if *Blue* retaliates by dropping *Blue's price* too, that triggers the two reinforcing loops in the top half of the diagram essentially simultaneously. The outcome could be that both *Red's customers* and *Blue's customers* wobble about a little as those customers who pay most attention to price comparison websites switch back and forth. But overall, the numbers of *customers* will remain about the same.

What will happen, though, is that *Red's profits*, and *Blue's profits* will both plummet. Until either *Red* or *Blue* goes out of business.

The player with the deepest pocket wins.

As noted on page 91, keeping all those competing reinforcing loops in a reasonably stable equilibrium is difficult – which is why the major players in such markets often agree that it's in their mutual self-interest to form a cartel. But if someone breaks ranks, that can trigger considerable instability – as exemplified by the oil price war between Saudi Arabia and Russia that broke out in March 2020, as will be discussed on pages 116 and 117.

The Investment 'Ladder'

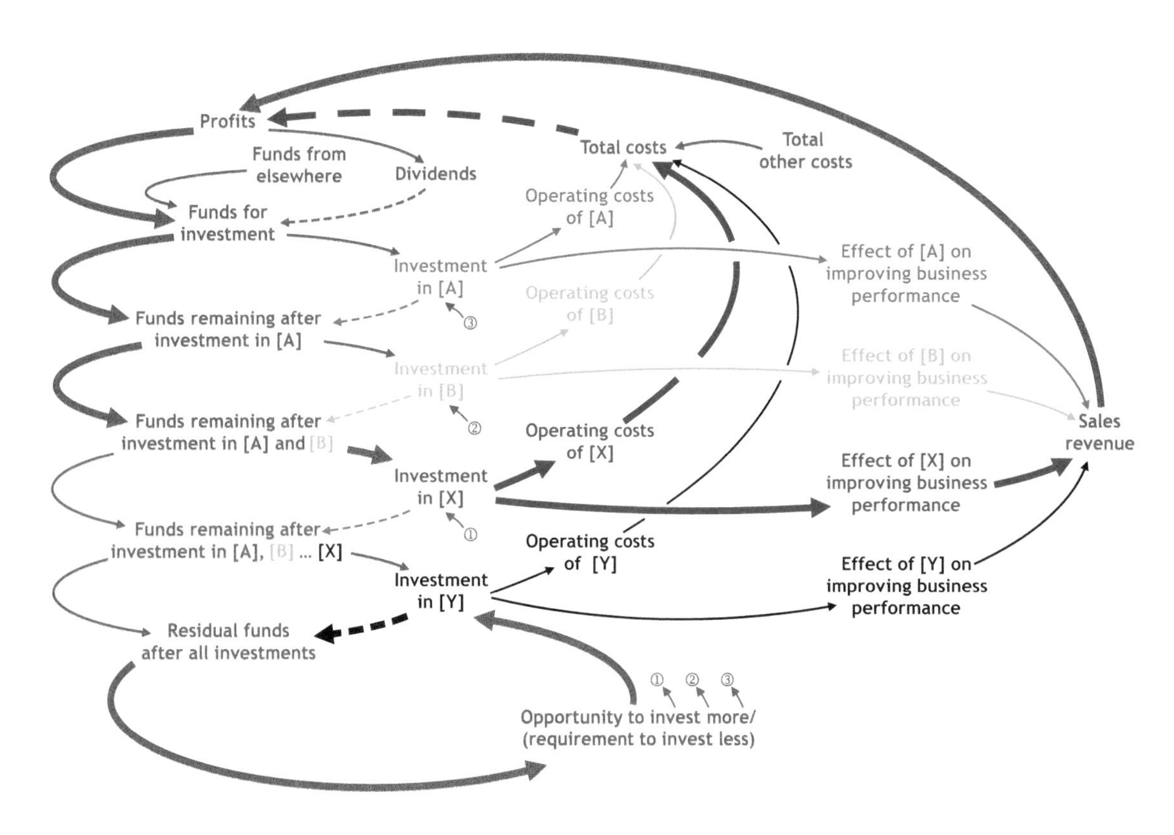

For narrative, see following page

The Investment 'Ladder'

In this story, both *Red* and *Blue* could choose to invest in, and compete on, *price* and *service*. In reality, any enterprise has many more opportunities for investment and competition, some – like *price*, *service* and product quality – being directly visible to the *customer*; others, such as training and a new accounting system, might be more behind-the-scenes. Ultimately, however, all investments must improve business performance – if any investment doesn't, why invest?

The causal loop diagram on page 94 extends the story of *Red* and *Blue* to a more general case, with *Red* having any number of possible investment opportunities *[A], [B], ... [X], [Y].*

Once *dividends* have been paid, *Red's funds available for investment* comprise the *profits*, less the *dividends*, plus any *funds from elsewhere*, such as additional share capital or borrowings. *Red* then decides to *invest in [A]*, and plans a corresponding allocation of funds, so determining the *funds available after investment in [A].*

The decision to *invest in [B]* determines the *funds available after investment in [A] and [B]* ... and so on for *investments in [X] and [Y].* Each *investment* drives two loops: one reinforcing, growing the business: the other balancing, increasing *costs* and *depleting* profits. All these loops operate simultaneously, some acting faster than others, some more effectively than others. The overall outcome is very hard to predict: does the business grow, or not?

Having planned to allocate funds to all investment opportunities, the resulting *residual* might be zero, implying that all the *funds available* have been used. If the resulting *residual* is positive, this sum might be transferred to reserves, or used to increase the investment in one or more of *[A], [B]* ... until the *residual* becomes zero – hence the balancing loops from *opportunity to invest more* to each of the *investment* opportunities.

And if the *residual* is negative, *Red's* current plans imply an over-spend, which might be financed, for example, by borrowing (as will be discussed on pages 112 and 113), or might imply a *requirement to spend less*, once again triggering the balancing loops to ensure that the total *investment* does not exceed the original *funds available*, whilst being sensibly balanced over the opportunities *[A], [B]* ...

This diagram assumes a management policy in which investment opportunities are ranked, so as to form a sequential 'ladder'. That of course is not the only possibility – perhaps the allocation decision is more of a 'free-for-all scramble', with the more powerful winning the larger slices of the cake. But as this diagram suggests, a much more rational approach is on the basis of each opportunity's *effect on improving business performance* (or, more generally, *enterprise performance*). Yes, these *effects* might not be easy to measure. But that doesn't stop them from being important to measure. And one of the benefits of systems thinking and of causal loop diagrams is the identification of what is truly important.

Chapter 8

Controlling Stock Levels

DOI: 10.4324/9781003304050-10

Controlling Stock Levels

Pat was doing the final assembly of a machine that was due to have its final quality control check after lunch, with the intention of despatching it later that afternoon.

When attaching a special cable to connect the motor, Pat noticed that the insulation was damaged. Pat decided it would be better to use an undamaged cable, and on checking the computer system, saw that there were two cables in stock.

'Sorry, Pat – we're out of stock on those,' said the stores clerk.

'But the computer said we're OK.'

'I'm sure it does. But we really don't have any. They're on order though – should have some next week.'

'That's no good – I need one right now!'

'Sorry…'

As Pat returned to his almost-assembled machine, he noticed another machine, that wasn't due for despatch until next week, which already had that cable connected.

'Ah. I think I've just solved my problem…'

Why We Hold Stocks

There are many situations in which we need to use something now – whether it's a specialised cable for a machine as in the story on page 98, or toothpaste this morning. Accordingly, we hold local stocks so that whatever we need is readily available for whenever that need arises. This happens along the entire supply chain: the local corner shop or supermarket holds stocks of toothpaste to meet the demand of all its expected shoppers, the regional warehouse sufficient stocks to supply the areas' shops, and so on back to the original source.

Holding stocks however, can be costly, for those items have to be paid for. Also, whilst 'stuff is on the shelf', there is a risk of deterioration (for example, for fresh foods), damage, obsolescence and theft. So, in general, stocks should be minimised, which can be a very complex problem-to-solve, since patterns of usage can be very irregular, and it takes time for items to be manufactured and distributed. And the development of 'just-in-time' supply chains, whereby intermediate stocks are very low, if not zero, so that every [this] arrives where it is needed, just as it is needed, is a tribute to the scientists, engineers and managers who have developed them. But it is a risk too, for if something goes wrong along that supply chain, then the consequences of shortages can be very damaging – as happened, for example, in March 2020 when a container ship ran aground in the Suez Canal, blocking one of the most vital of the world's supply routes, and disrupting global trade.

Given that supply chains are, by necessity, 'joined up', they are very well-suited to being analysed and understood using systems thinking. The following pages therefore 'open the door' by looking at a simple, but realistic, example.

When Stock Is Available…

Suppose that there are, say, 20 units of a particular item *available* at a particular time. If the immediate *requirement* for that item is, say, 4 units, then that *requirement* is *satisfied*, and the *consumption* of those 4 units depletes the *available stock* to 16 units. This can be represented as a closed feedback loop containing a single inverse link, which is therefore a balancing loop.

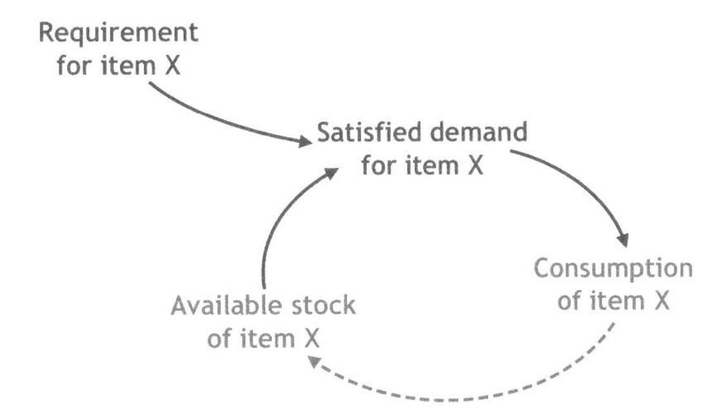

But although this loop looks very similar to the 'target' balancing loop shown on page 68, this balancing loop does not converge on the *requirement for item X*, for the *satisfied demand* is not a 'gap', a difference between a 'target' and an 'actual'. Rather, as will be shown on the following pages, the balancing loop shown here is embedded within a more complex system, and so its behaviour needs to be examined within a wider context.

Successive *satisfied requirements* progressively deplete the *available stock*, and if nothing else happens, a point is reached at which the *available stock* (say, 3 units) is less than the current requirement (say, 5 units), implying that the *demand cannot be fully satisfied*. What happens then is shown on the next page.

...and When It Isn't

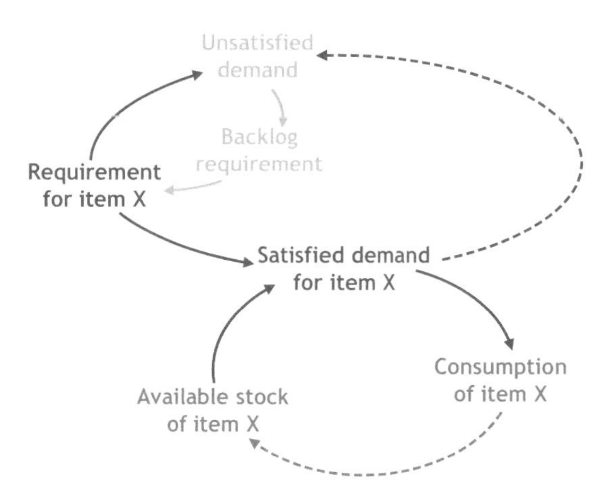

If there are 3 units remaining in the *available stock*, a *requirement* for 5 units can be partially satisfied by reducing that *stock* to zero. At the same time, there is an *unsatisfied demand* of 2 units, this being the *requirement* less any *satisfied demand*. The link from *requirement* to *unsatisfied demand* is therefore direct, and the link from *satisfied demand* to *unsatisfied demand* is inverse.

The *unsatisfied demand*, and the resulting *backlog*, adds to the *requirement* that needs to be *satisfied* at some time in the future, and until the *available stock* is replenished, any future new *requirements* will remain *unsatisfied*, making the *backlog* even bigger.

At any time, the *satisfied demand* is the lesser of the immediate *requirement* and the current *available stock*. In mathematical terms, this can be expressed using the MINIMUM function as

Satisfied demand = MIN (*Requirement, Available stock*)

Importantly, the *satisfied demand* is NOT the difference between the *requirement* and the *available stock*; rather, it is either the one, or the other, depending on which is the lesser. The links from *requirement* to *satisfied demand*, and from *available stock* to *satisfied demand*, are therefore both direct: for any *requirement* less than a given *available stock*, the greater the *requirement*, the greater the *satisfied demand*; likewise, the greater the *available stock*, the greater the likelihood that any *requirement* will be *satisfied*.

The Right Way to Meet Unsatisfied Demand...

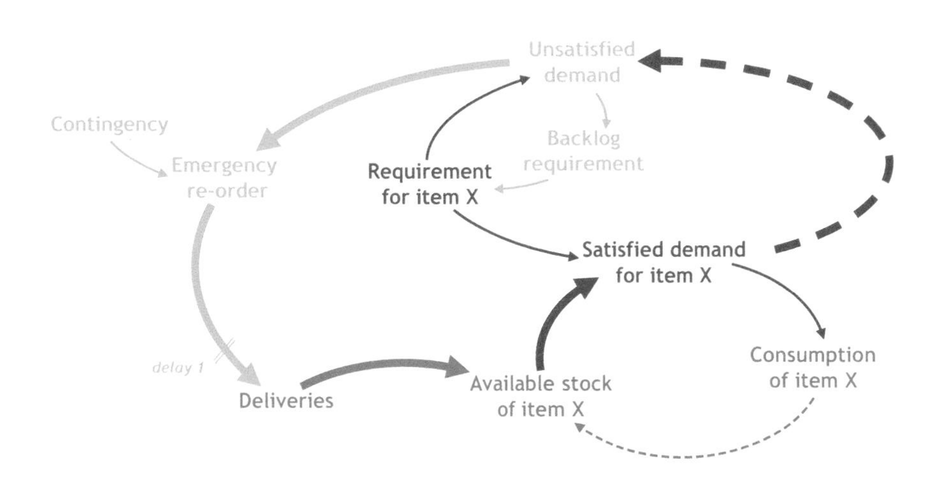

One consequence of any *unsatisfied demand* is to place an *emergency* re-order with a supplier, perhaps with some added contingency – or to run to the local shop for that toothpaste!

The *emergency re-order* will – after a possible, and hopefully short, *delay* – result in a *delivery,* so replenishing the *available stock*, enabling the current *requirement*, and any associated *backlog*, to be satisfied. That assumes, of course, that the quantity associated with the *emergency re-order* was sufficient, and that the supplier of that *re-order* themselves had that stock available.

The highlighted closed feedback loop contains a single inverse link, and so is a balancing loop...

...but unlike the *consumption* balancing loop, this *emergency re-order* balancing loop does act to converge on a 'target' and to control the system to ensure that the *available stock* can meet the *requirement*. In this loop, the *unsatisfied demand* is the difference between the *requirement* and the *satisfied demand*, and so acts as a 'gap'. When the *unsatisfied demand* is a positive number, this triggers an *emergency re-order*; when the *unsatisfied demand* becomes zero, no more *emergency re-orders* are triggered, and the loop 'switches off'.

...and the Wrong Way

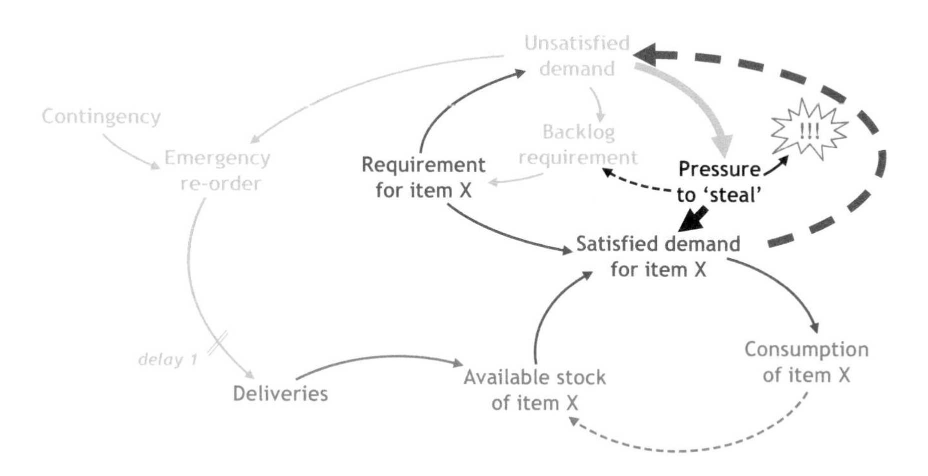

An alternative response to any *unsatisfied demand* – as was hinted at in the story on page 98 – is to *steal* what you need from somewhere else. That immediately *satisfies the demand*, as indicated by the highlighted balancing loop which brings the *unsatisfied demand* back to zero.

Stealing, though, can have nasty consequences. I won't get away with *stealing* my wife's tube of toothpaste (although my inner demon tells me that she might not notice just a small squeeze...); and certainly, in an industrial context, *'stealing'* components from other assemblies can create huge problems. So it really should not be done. But we can all recognise the temptation...

Re-Ordering...

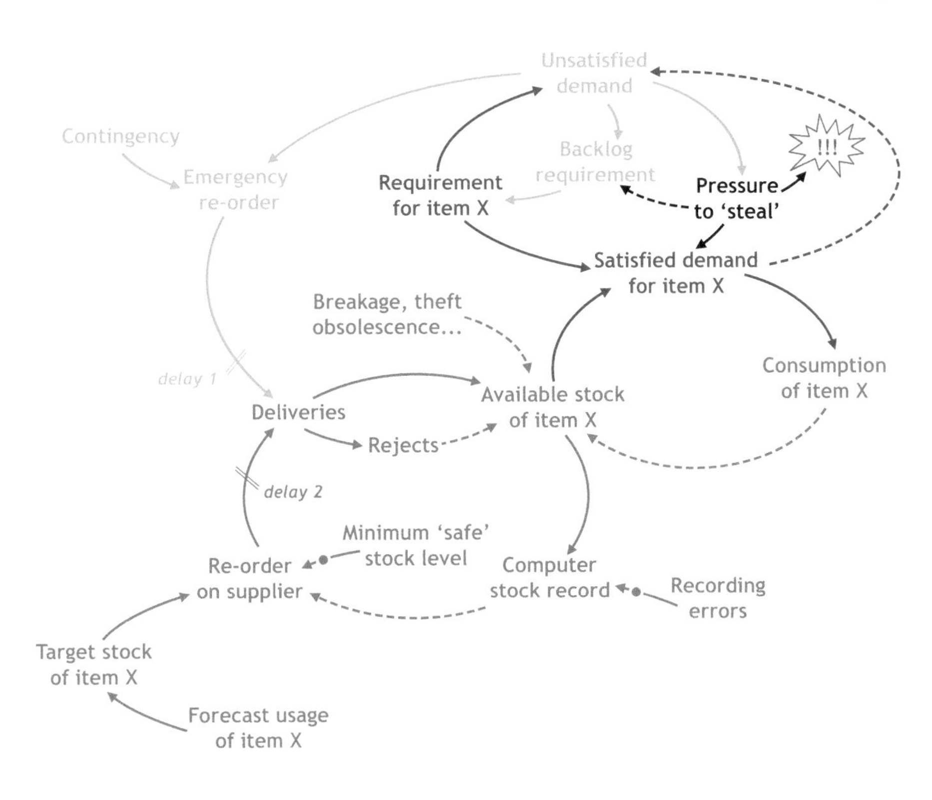

Unsatisfied demand should be an exceptional, and rare, event: the norm is to *re-order* an appropriate quantity before the *available stock* becomes too low.

Accordingly, the *available stock* is continuously monitored, for example, by using a *computer system*. Whenever the *stock record* indicates that the *available stock* has reached, or just fallen below, a pre-determined *minimum 'safe' stock level*, this triggers a *re-order on the supplier*, for an appropriate quantity, such as the *target stock anticipated to be needed in the future*, less the *(computer record) of the available stock*.

Note that the link from *minimum 'safe' stock level* to *re-order on supplier* is an influence link. That's because the *'safe' level* does not determine the quantity re-ordered (that's the *target stock*); rather, the *'safe' level* determines when the re-order is made: as soon as the *stock record* falls below the *'safe' level*.

The *available stock – computer record – re-order – deliveries* loop contains one inverse link, and so is a balancing loop acting to bring the *available stock* into line with the *target stock*, and – as suggested in this diagram – without any delay. Some supply chains work very quickly, others less so, as will be discussed on the next page.

Also shown are some of the 'real-life' events that also need to be recognised, such as *items rejected on receipt from the supplier*, perhaps as a result of a quality problem; *breakage, theft* and *obsolescence*; and also *errors in the computer record* – perhaps because items have been *consumed*, but without being *recorded*. Also, two key parameters in this diagram are the *minimum 'safe' stock level* and the *forecast usage*, both of which are shown as dangles, but that is a simplification: these themselves can be determined from other factors such as the lead time for delivery, and patterns of past, and possible future, *consumption*.

...with a Delay

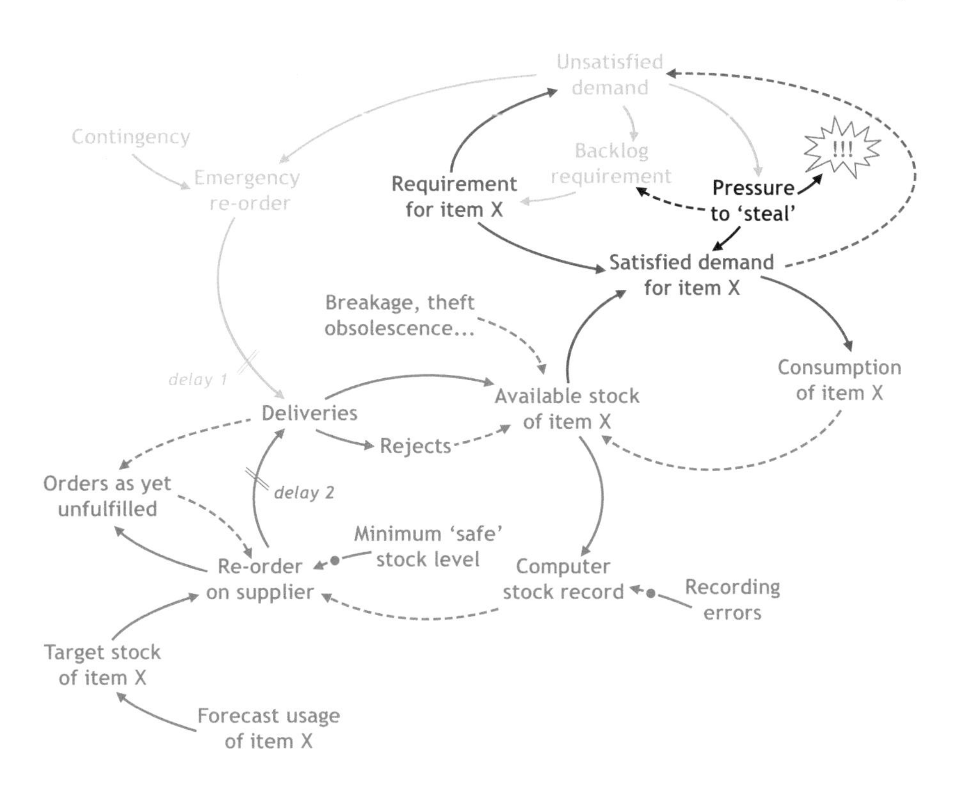

In many supply chains, there can be a delay between the placing of an order and the subsequent delivery.

During that time, the *available stock* continues to be *consumed*, and to be below the *minimum 'safe' stock level*. In principle, that would trigger a series of *re-orders*, which would in practice be erroneous, for the *minimum 'safe' stock level* should have been set recognising that there is an *order delay*, and anticipating that the *available stock* would reduce below the *minimum 'safe' stock level*, but with a low probability of running out and so causing *unsatisfied demand*.

The fact that an *order has been placed but not yet fulfilled* therefore needs to be taken into account, acting to prevent further, unnecessary *orders* from being triggered.

Accordingly, a *re-order* triggers a *delivery*, and also registers as an *order as yet unfulfilled*. When the *order*, after the *delay*, is *delivered*, the *orders as yet unfulfilled* are reduced by the appropriate quantity – hence the inverse link from *deliveries* to *orders as yet unfulfilled*. But whilst *delivery* of the *order* is awaited, the quantity associated with *orders as yet unfulfilled* is subtracted from what an *order* might otherwise have been, reducing the corresponding quantity to zero – hence the inverse link from *orders as yet unfulfilled* to *re-order on supplier*. This explicit recognition of the *delay*, and of *orders as yet unfulfilled*, therefore eliminates the possibility of unnecessary *re-orders*.

Also, and not explicitly shown on the diagram, note that the *consumption* over the time during which *orders* are awaited might influence both the *re-order quantity* and the *minimum 'safe' stock level*.

Chapter 9

Queues, Angry Customers, Borrowing, Supply and Demand

DOI: 10.4324/9781003304050-11

Queues and Angry Customers

BARGAIN TEDDY BEAR OFFER SPARKS VIOLENT SCENES ACROSS BRITAIN

A marketing ploy to get more shoppers into a toy store had to be abandoned after it caused huge queues and outbreaks of violence across Britain.

Build-A-Bear Workshop was offering UK customers the chance to buy any bear, for the price of their child's age. Huge demand caused chaos at shops across the UK, with some forced to close their doors.

Children were left distraught after queuing for hours only to be told the bears had run out, and in other places there were reports of fighting between parents desperate to purchase one of the bears.

The Guardian, 12 July 2018

https://www.theguardian.com/uk-news/2018/jul/12/build-bear-workshop-bargain-teddy-offer-sparks-chaotic-scenes-across-britain.

We all experience queues, sometimes with patience, sometimes with irritation, perhaps sometimes even with anger. This chapter explores the systemic behaviour of queues, and two rather different, but structurally related, themes – borrowing and prices.

Queues and Angry Customers

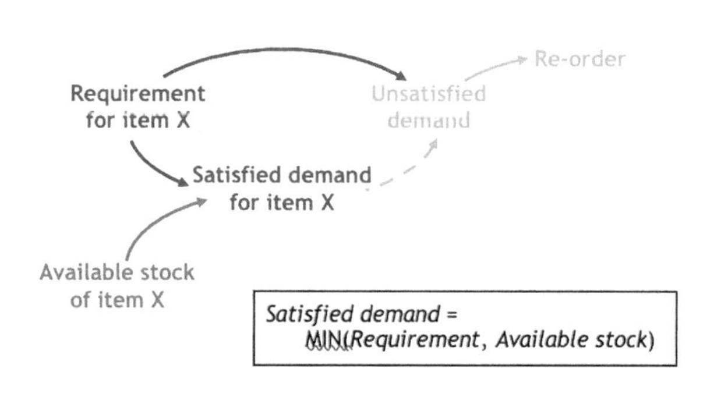

As we saw on page 101, the *satisfied demand for a stock item* is the lesser of the *requirement for that item*, and the *available stock*. This is shown in the diagram to the left, in which the *satisfied demand* is represented mathematically using the MINIMUM function.

The *unsatisfied demand* resulting from running out of *stock* is just one example of a much more general context in which a constraint (in this case, the *available stock*) limits the successful delivery of a required outcome (in this case, the satisfaction of the total *requirement for item X*).

The purpose of this chapter is to present some further examples, noting that these are not 'the whole story', but rather represent 'building blocks' that would feature within more complete causal loop diagrams.

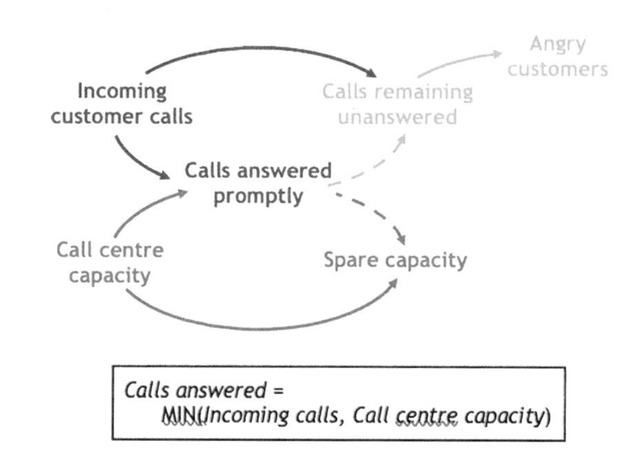

Another example in which a constraint limits delivery is what happens when the number of *incoming customer calls* to a call centre exceeds the *call centre capacity*. As shown in the diagram to the right, the *unanswered calls* end up in a queue of increasingly *angry customers*, some of whom may not wish to remain *customers* for much longer.

In many organisations, any *spare capacity* is regarded as 'waste', and so the actual *capacity* is planned to be as low as the organisation's management believe they can get away with without making too many *customers* upset. That makes the delivery system very vulnerable to upsurges in demand, as the National Health Service in England discovered during the Covid-19 crisis.

Hospital Admissions

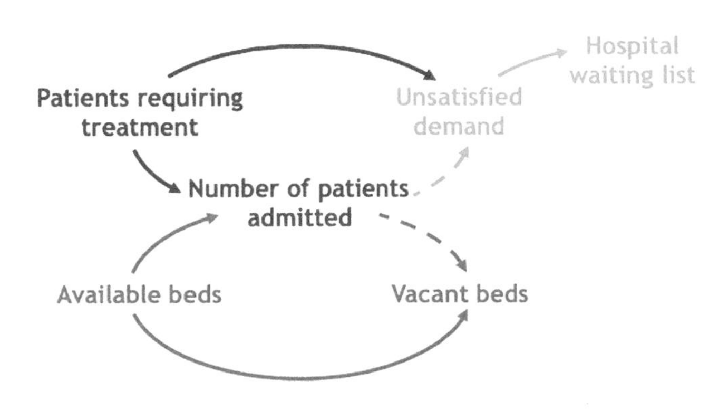

Patients admitted = MIN(Patients requiring treatment, Available beds)

This diagram relates to hospital admissions: the *number of patients admitted* over any time period is the lesser of the number of *patients requiring treatment* and the number of *available beds*. The *unsatisfied demand* is the excess of the *patients requiring treatment* over the *number of patients admitted*, so adding to the *waiting list*. The number of *vacant beds* – if any – is the difference between the *number of patients admitted* and the number of *available beds*.

However, as noted on the previous page, this diagram does not tell 'the whole story', nor does that on page 46; both are fragments of a richer, more complex, diagram...

Hospital Admissions – A More Realistic View

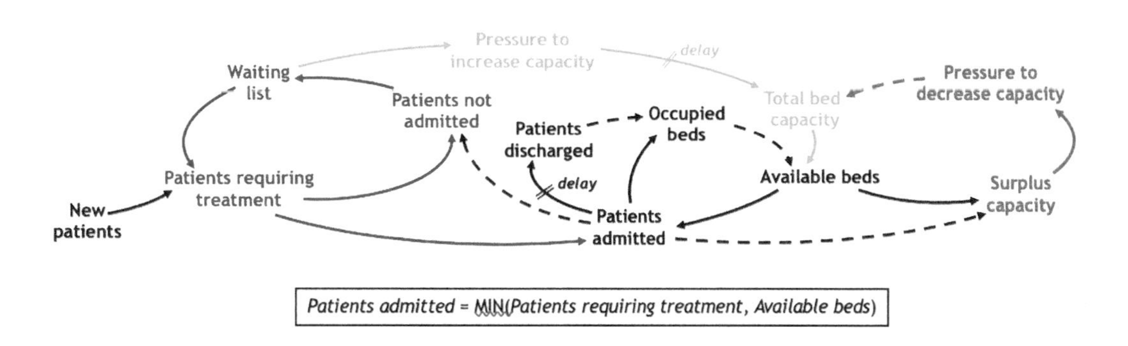

This causal loop diagram builds on those shown on pages 46 and 110, and adds some further, more realistic, features.

The number of *patients admitted* each day is expressed, as is now familiar, as MIN (*Patients requiring treatment, Available beds*); the number of *patients not admitted* (this being the unsatisfied demand) is the difference between the number of *patients requiring treatment* and the number of *patients admitted*; and the *surplus capacity* is the difference between the number of *available beds* and the number of *patients admitted*. Also, *admissions* cause *beds to be occupied*, so reducing *availability*, until the *patient is discharged* after a *delay* corresponding to the time required for the patient's treatment and recovery in the hospital.

The number of *patients not admitted* on any day adds to the *waiting list*, resulting in progressively stronger *pressure to increase capacity*. This introduces a balancing loop bringing the *total bed capacity* up to meet the higher number of *patients requiring treatment*. In practice, however, it can take a considerable time to build the required infrastructure.

If, however, the number of *patients requiring treatment* steadily falls, and there is *surplus capacity*, this can exert *pressure to decrease the capacity* accordingly, as shown by the balancing loop to the right.

Overall, this system – in principle – continually adjusts the *total bed capacity* to the number of *patients requiring treatment*; in practice, *delays* can cause a relatively sluggish response, especially to increases in *capacity*.

Funding Investments by Raising Share Capital or Borrowing

Pages 94 and 95 discussed how investments can drive business growth, and the assumption was made that the investments are all funded by, and constrained by, the business's profits. Another possibility is to raise additional funds, either as *new share capital* or by *new borrowing*, as illustrated by the causal loop diagram on page 113 and discussed here.

The *investment funds allocated* are the lesser of the *requirement for investment*, and the *investment funds available* from within the business: *funds allocated* = MIN (*Requirement*, *Funds available*). If the *funds available* are greater than the *funds required*, then all *requirements* are *funded*, and any *surplus* can fund appropriate *additional investments* up to the point at which all the *investment funds available* have been fully used up...

...as represented by the balancing loop connecting the *requirement for investment*, *investment funds allocated*, *surplus investment funds available* and *additional investments*, which acts to bring the *requirement for investment* (as enhanced by the *additional investments*) into line with the *investment funds available*.

If, however, the *requirement for investment* exceeds the *investment funds available*, the *investments* could be rationed as discussed on pages 94 and 95, or the *requirement for additional funds* might be met by raising *new share capital* or by *new borrowing*. In the causal loop diagram on page 113, *new borrowing* is the preference, with *new share capital* determined as the difference (*requirement for additional funds* – *requirement for new borrowing*); if any *new share capital* is decided upon first, then the inverse link is from *new share capital* to *requirement for new borrowing*.

It is possible that the organisation already has some *existing borrowing*, and there also might be a *borrowing cap* limiting the maximum total amount that the organisation might *borrow* at any time. The *maximum borrowing available* is therefore the difference between the *borrowing cap* and the *existing borrowing*.

The *maximum borrowing available* may now be compared to the *requirement for new borrowing*, and the *actual new borrowing* is the lesser of these two figures. The *actual new borrowing*, and any *new share capital*, together fund the resulting actual *additional investment*. If this is insufficient to meet the original *requirement for additional funds*, there is a *remaining unsatisfied funding requirement*, and the corresponding investment has to be limited accordingly.

Also, if the *actual new borrowing* is less than the *borrowing available*, then some *borrowing remains available* within the overall *borrowing cap*. The *remaining available borrowing* is therefore the '*existing borrowing*' as at any time after the *actual new borrowing* has been drawn down, and so the *remaining available borrowing* replaces the *existing borrowing* – in accounting terms, the *remaining available borrowing* is the 'closing balance' after the transaction represented by the *actual new borrowing* has been deducted from the 'opening balance', the original *existing borrowing*.

Borrowing

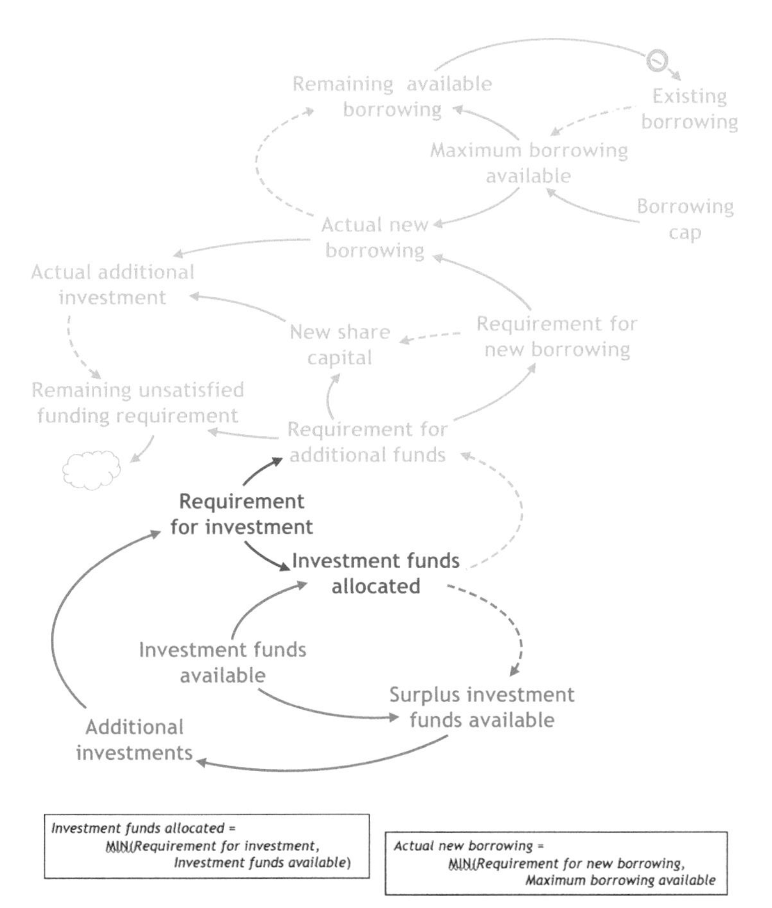

In this diagram, two variables are calculated using the MINIMUM function.

First, the *investment funds allocated* are the lesser of the *requirement for investment* and the *investment funds available*...

...and second, the *actual new borrowing* is the lesser of the *requirement for new borrowing* and the *maximum borrowing available.*

This diagram also introduces a new feature – a 'special' form of direct link, from *remaining available borrowing* to *existing borrowing*, as shown by the circle close to the 'pointy' end of the arrow. This is known as a **replacement direct link**, for its action is for the variable at the 'blunt' end of the arrow (in this case, the *remaining available borrowing*) to replace the variable at the 'pointy' end of the arrow (in this case, the *existing borrowing*) at the start of the next time period, just as an accountant's 'closing balance' at the end of any time period becomes the next time period's 'opening balance'.

The 'cloud' represents the consequences of any *remaining unsatisfied funding requirement.*

Supply and Demand

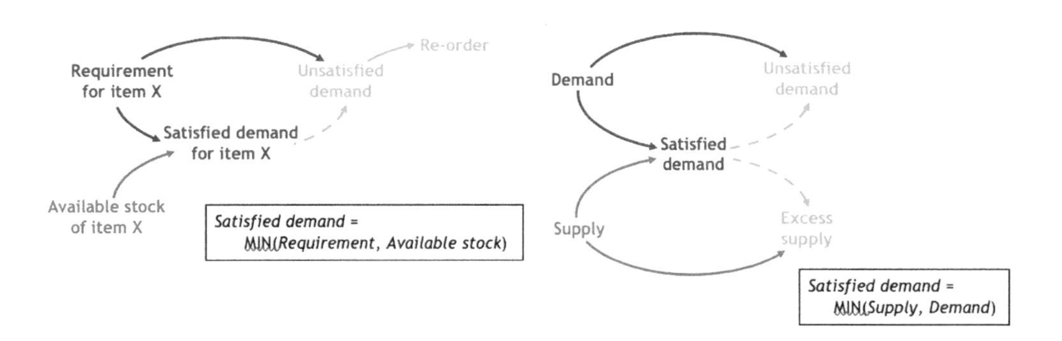

As we saw on pages 101 and 109, the diagram to the left represents what happens when the *requirement for item X* exceeds the *available stock*, as exemplified by stocks held locally in a factory's stock room. The *satisfied demand*=MIN (*Requirement, Available stock*) is limited by the *available stock*, resulting in an *unsatisfied demand*=*requirement*–*available stock*.

This may be generalised to illustrate the matching, or otherwise, of the supply of any product or service (shown in the above diagram as the *available stock*) and the corresponding *demand* (the *requirement*), as shown to the right.

The *satisfied demand*, as we have already seen, is the lesser of the *supply* and the *demand*; the *unsatisfied demand* is, as before, the difference between the *demand* and the *satisfied demand*; and the *excess supply* is the difference between the *supply* and the *satisfied demand*:

Satisfied demand = MIN *(Supply, Demand)*
Unsatisfied demand = *Demand* – *Satisfied demand*
Excess supply = *Supply* – *Satisfied demand*

The diagram to the right is fundamental in understanding the nature of prices, and the behaviour of commercial markets, as discussed in detail in the next chapter.

Chapter 10

Prices, Inflation, Economic Depression and Growth

DOI: 10.4324/9781003304050-12

The Saudi Arabia – Russia Oil Price War, March 2020

OIL SLUMPS AS SAUDI ARABIA LAUNCHES PRICE WAR

The oil price has suffered an astonishing plunge overnight.

Brent crude slumped by 30% at the start of trading, after Saudi Arabia effectively launched an oil price war against competitors such as Russia and the US.

After failing to get Russia's agreement for major supply cuts, Saudi authorities retaliated by slashing export prices and boosting production.

This sent Brent crude reeling to just $31 per barrel at one point – its lowest since 2016 – down from $45 on Friday night.

The Guardian, 9 March 2020

https://www.theguardian.com/business/live/2020/mar/09/markets-plunge-crash-financial-crisis-coronavirus-ftse-italy-oil-price-dow-business-live?page=with:block-5e65e9ee8f085f0b8d944bc3&filterKeyEvents=false - liveblog-navigation.

As this story illustrates, supply, demand and price are mutually inter-related…

A (Very Simplified!) Description of the 2020 Oil Price War

One of the consequences of the Covid-19 pandemic was to slow down the entire world economy, and in January and February 2020, this caused a drop in the demand for oil, pushing the oil price down. If the supply of oil were to continue at the late 2019 levels, the price would continue to fall, which would not be good for the major oil producers, such as Saudi Arabia and Russia.

Saudi Arabia therefore sought agreement across all producers, including Russia, to cut production, with the intention of keeping supply and demand at similar levels, so holding the price up.

Russia, however, refused.

In response, on 7 March, Saudi Arabia announced a deep cut in their oil price; very soon thereafter, they stated their intention to increase production, which happened in April; they subsequently cut production substantially back in May and June.

Since Saudi Arabia controls so much of the oil market, these actions caused the general market price – as expressed by the 'Brent crude oil price' – to fall, and then recover, as shown in the chart to the right…

Source: https://www.macrotrends.net/2480/brent-crude-oil-prices-10-year-daily-ch

…but it was not until late 2020 that the supply and demand for oil had both adjusted to restore a stable price – albeit at a level somewhat lower than the price had been before Saudi Arabia's March cut.

This is just one example of how a 'big player' in any market can force a drop in price, usually with the dual objectives of capturing more market share and causing a competitor to incur unsustainable financial losses, ultimately driving that competitor out of business. Even though the 'big player' is likely to make losses too, if it has deep enough pockets and can fund those losses without becoming bankrupt, as soon as the competitor has dropped out of the market, the survivor can then raise the price, and benefit from the increased market share.

In all markets, of which the international oil market is just one example, supply, demand and price are closely interlinked. The following pages therefore explore the systemic interactions between supply, demand and price, as represented using causal loop diagrams.

Supply, Demand and Price...

For any product or service, the *satisfied demand* at any given *price* is the lesser of the *supply* and the *demand*. As we saw on page 114, this is another example of a MINIMUM function: *satisfied demand* = MIN(*supply, demand*).

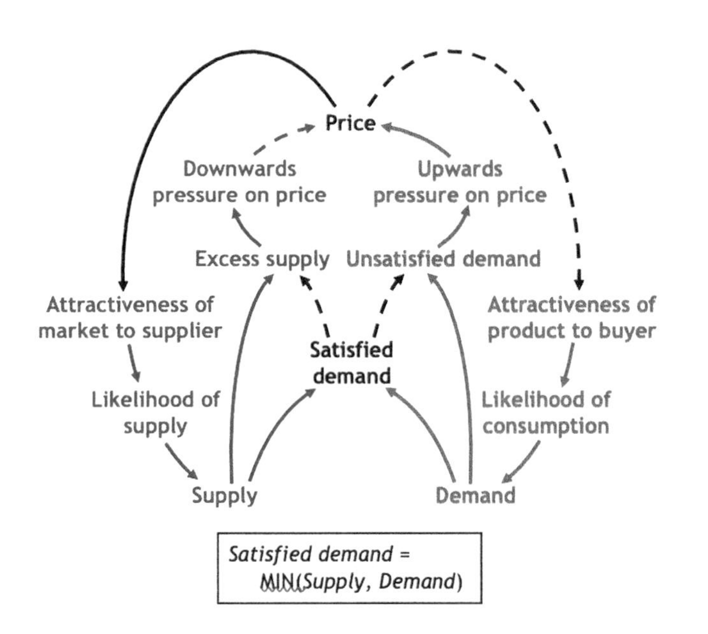

Any imbalance between *supply* and *demand* therefore results in either *excess supply* (when *supply* > *demand*) or *unsatisfied demand* (when *supply* < *demand*).

In a free market, *supply, demand* and *price* can all vary, and so this chapter examines the behaviour of the *supply, demand, price* system. Our first analysis is what happens when *demand* suddenly falls, resulting in *excess supply*. In many markets, suppliers who think they are going to be left with unsold stock, which might perhaps soon become obsolete, are likely to be more willing to accept a lower *price*, so putting *downwards pressure on the price.*

A lower *price* increases the *attractiveness of the product to the buyer,* which, in turn, increases the *likelihood of consumption,* thereby increasing *demand,* as, for example happens when shops have 'sales'. This increases the *satisfied demand*, so reducing the *excess supply* which started the process off.

The initial decrease in *price* has an effect on the *supply* side too. Decreasing *price* brings the threat of reduced profits, making the *market increasingly less attractive to suppliers*, so reducing *the likelihood of supply*. Over time, the resulting lower *supply* also serves to reduce the *excess supply*.

Whilst there is *excess supply*, the *satisfied demand* equals the *demand,* and so the value of the *unsatisfied demand* is zero. This implies that the *upwards pressure on price* is also zero. The continuing *excess supply* causes the *price* to continue to fall, driving a decrease in *supply* and a simultaneous increase in *demand*. Together, these cause a decrease in the *excess supply* until this becomes zero, at which point the *downwards pressure on price* becomes zero and the *price* stabilises. This re-establishes the *supply–demand* balance, with all *demand* satisfied at the appropriate *price* – as illustrated by the simulations whose results are shown on pages 122–125.

This page has examined how a market responds to excess supply; the next page explores the alternative, unsatisfied demand, in which demand exceeds supply, and buyers are unable to purchase as much as they wish to.

...and Some More on Supply, Demand and Price

When d*emand is not fully satisfied*, or there is fear of this, potential buyers are usually willing to pay more to secure whatever they can of the limited *supply*. This *upwards pressure on the price* is sensed by the suppliers, who increase the *price* accordingly.

On the *demand* side, an increase in *price* makes the *product less attractive,* so driving down *demand*, and reducing the *unsatisfied demand*; on the *supply* side, as the *price* increases, the prospects of good future profits are raised, making the market *more attractive to suppliers*. More new product enters the market, and so *supply* increases, thereby increasing the *satisfied demand*, and decreasing the *unsatisfied demand*. By triggering a rise in *price, unsatisfied demand* both decreases *demand* and increases *supply*, so restoring the *supply*: *demand* balance and eliminating the original *unsatisfied demand*. And throughout the time for which there is *unsatisfied demand, supply* equals *satisfied demand,* and so the value of the *excess supply* is zero. This implies that the *downwards pressure on price* is also zero.

The causal loop diagram on the previous page captures these behaviours. Structurally, this system comprises two pairs of balancing loops which collectively act to stabilise the whole system, as will be discussed in more detail on the following two pages. An important feature of this system – as is evident from the causal loop diagram shown in page 118 – is that there are no dangles and no targets: the system achieves equilibrium 'all by itself'. Furthermore, changes to any one of the three key variables *supply* (for example, a strike in a factory causes a shortage), *demand* (a product goes 'viral' on the web, so driving a surge) or *price* (as happened in the oil market in March 2020 as described on pages 116 and 117) will trigger changes to the other variables, as the system responds to whatever shock takes place.

Two further features – or rather missing features – of this diagram are important. First, there is no mention of supplier costs, which are relevant in that an increase in supplier costs can exert an *upwards pressure on price*. The effects of supplier costs will be explored further on pages 129–133; for the moment, the assumption is made that supplier costs remain constant.

Second, the diagram on page 118 shows no *delays,* and assumes that 'signals' such as an initial *unsatisfied demand* or *excess supply* are detected, and acted upon, at once by a corresponding change in *price*. In reality, *delays* are inevitable, if only in the time taken for suppliers to change their *supply* schedules. And as we saw on page 71, *delays* in balancing loops can cause some quite erratic results. So the reality is much more complex – but understanding diagrams such as the one shown on the previous page, and enhancing them to incorporate, for example, the likely effect of delays, can be very helpful – as indeed can system dynamics simulation modelling, using the causal loop diagram as the specification for the computer simulation model, as shown in the examples on pages 122–125.

Excess Supply – Two Balancing Loops

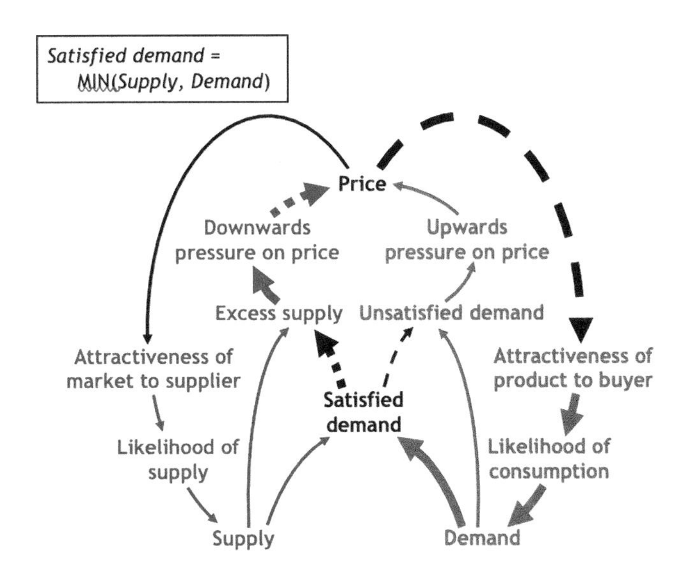

Excess supply can arise in two ways. First, as a result for example, of over-stocking, or a supplier seeking to 'dump' products into a market; second, from a fall in *demand* as driven by events 116 such as the depression of economic activity caused by the Covid-19 pandemic, which decreased the *demand* for oil, as described on page 117.

Excess supply puts *downwards pressure on price*, simultaneously triggering the *demand* side balancing loop (left) and also the *supply* side balancing loop (below), both of which act to bring the *supply* into line with the (now reduced) level of *demand* at a lower *price*.

The description of the behaviour of this system has focused on the two balancing loops. But there are some reinforcing loops too – for example, in the diagram to the right on the *supply* side, the loop from *supply* to *satisfied demand* to *excess supply*, and then following the emphasised links back to *supply*.

Although this reinforcing loop exists, it is never invoked: an increase or decrease in *supply* impacts the *excess supply* wholly through the direct link, and not along the path through *satisfied demand*.

This emphasises the importance of interpreting all causal loop diagrams in the context of real situations, and not as theoretical abstractions.

Unsatisfied Demand – Two More Balancing Loops

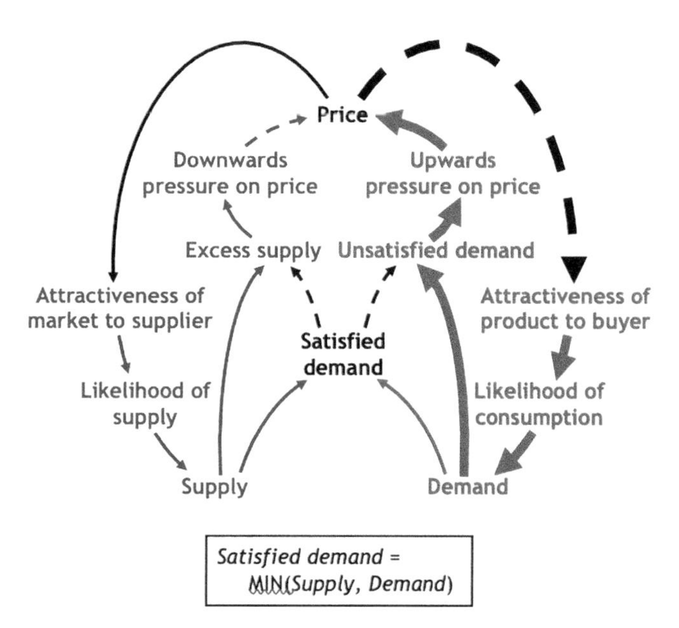

Unsatisfied demand can also arise in two ways: *supply* might decrease relative to (current) *demand* as the result of, for example, a strike that disrupts production or distribution; alternatively, *demand* might increase relative to (current) *supply*, as happened when the Covid-19 outbreak drove a sudden surge in the *demand* for face masks.

The resulting *upwards pressure on price* acts to diminish the *attractiveness of the item to the buyer*, whilst simultaneously *attracting more supply* into the market, so triggering the balancing loops on both the *demand* side (left), and also the *supply* side (below).

Both balancing loops operate together, and act to restore *supply:demand* equilibrium at the appropriate *price*.

And whilst there is *unsatisfied demand*, the value of the *excess supply*, is necessarily zero: there is therefore no *downwards pressure on price*.

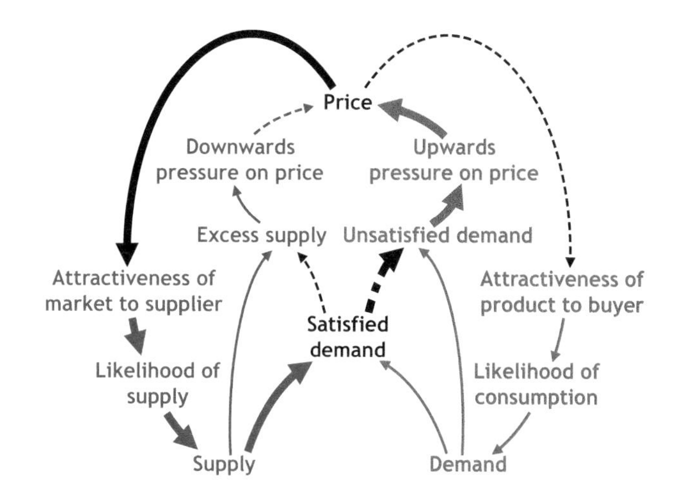

A Simulation of Supply, Demand and Price...

These graphs are the results of my computer simulation of the causal loop diagram shown on page 118, under various sets of assumptions that reflect the 2020 Saudi Arabia–Russia oil price war, as described on pages 116 and 117.

These two charts show what happens when *supply* quickly and smoothly adjusts to a sharp drop in *demand*. As described on pages 118 and 120, the corresponding *excess supply* causes the *price* to fall, and *supply*, *demand* and *price* all soon stabilise.

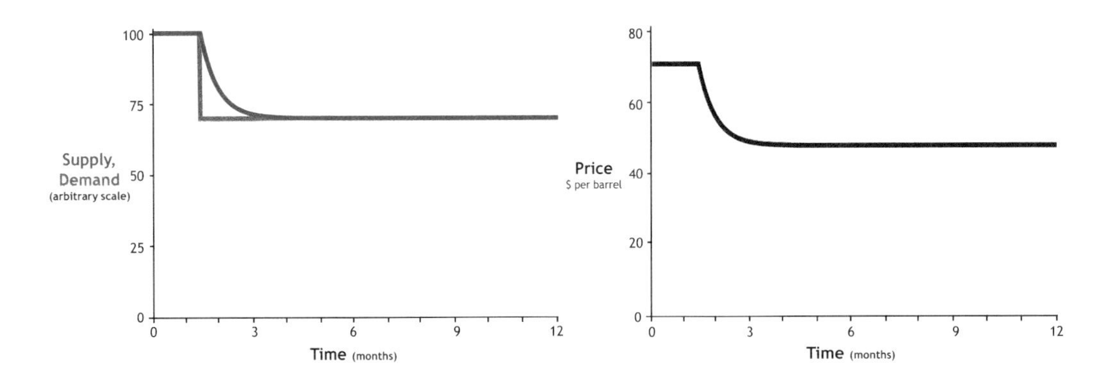

This is what Saudi Arabia wanted to happen as a result of the drop in *demand* triggered by the Covid-19 pandemic in early 2020. But that required all the major oil producers to cut back on *supply* together, which Russia was unwilling to do.

...and Another Simulation...

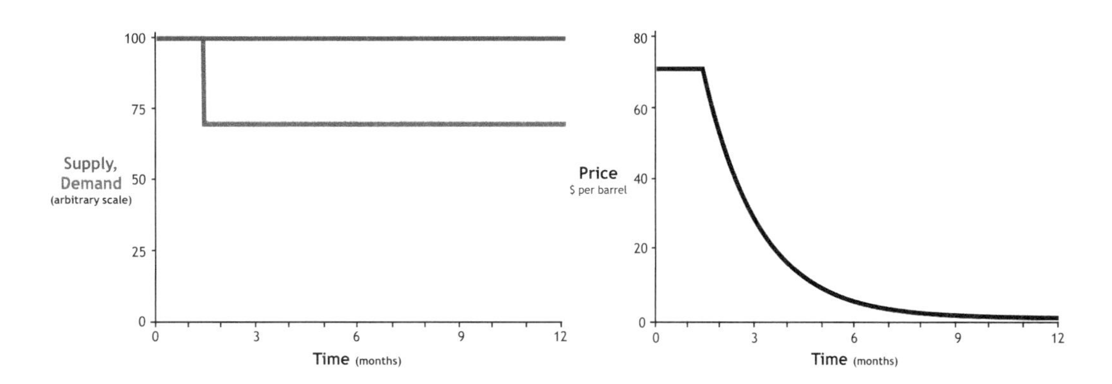

These charts show what Saudi Arabia wished to avoid: if *supply* remains constant even though *demand* has fallen, and remains at a lower level, then the *price* can fall precipitously, and to a level much lower than if *supply* were to adjust.

...and of the 2020 Saudi Arabia – Russia Oil Price War...

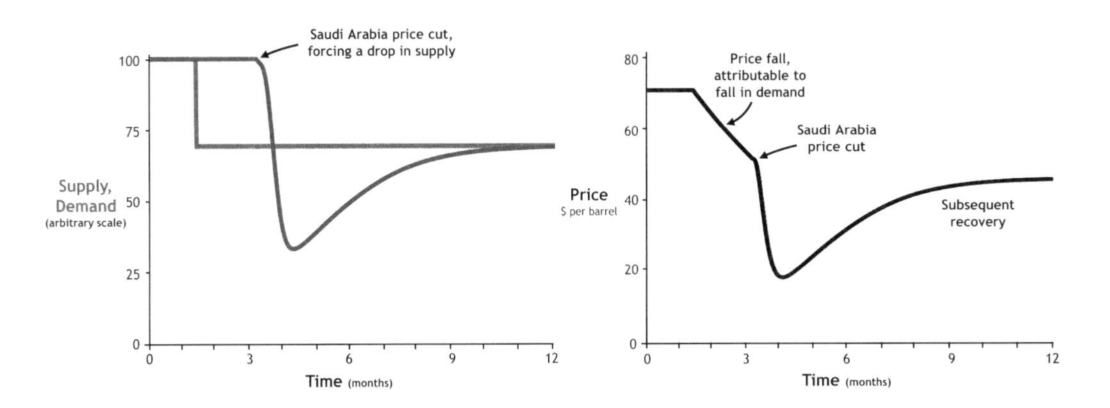

Initially, *supply* matches *demand*, and the *price* is stable. Then as a result of the pandemic, *demand* drops, but *supply* remains constant. The excess of *supply* over *demand* causes the *price* to fall.

Russia's refusal to reduce *supply* leads Saudi Arabia to drop its *price*, which affects the market as a whole. In this simulation, I have made the simplifying assumptions that this causes *supply* to fall, and then gradually to readjust, whilst *demand* stays constant. Over time, the *price* returns to level somewhat lower than that at the time of the cut.

The real events were of course far more complex than this simple model reflects; that said, the behaviour of the simulated *price*, as shown on the right, is sensibly similar to the actual behaviour as shown on page 117.

...and Some Further Simulations

The simulation results shown on pages 122–124 all assumed that the *demand* stayed constant after the initial downwards shock, and that only *supply* and *price* could respond. This (rather unrealistic) assumption is relaxed in the results shown on this page, in which *supply, demand* and *price* can all mutually adjust in response to an initial downwards shock in *supply* – a shock that results in an *unsatisfied demand*, as discussed on page 121.

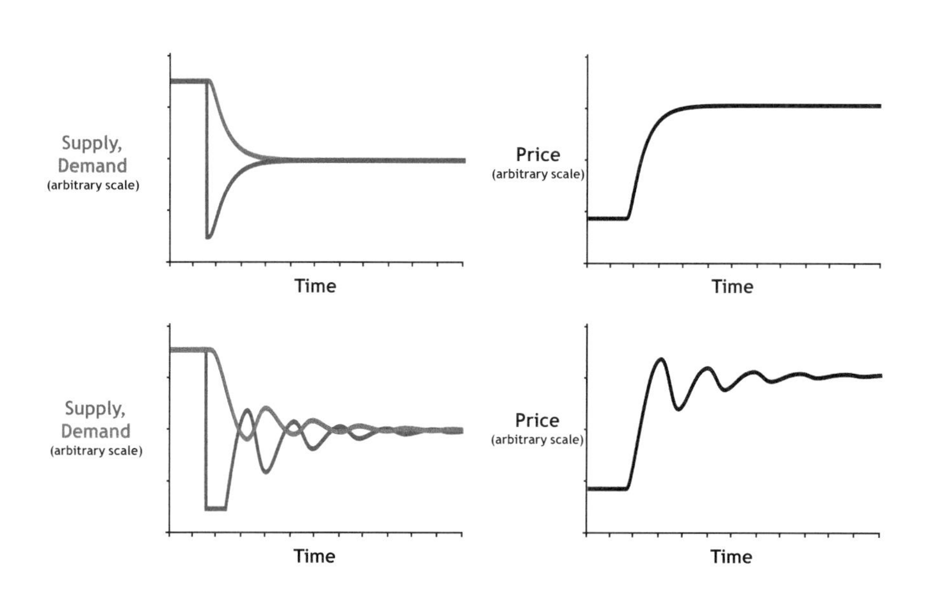

In the simulation whose results are shown in the two upper charts to the left, *supply, demand* and *price* all respond quickly and smoothly to stabilise at a lower *supply* and *demand*, at a correspondingly higher *price*.

But in the two lower charts, various delays, such as a lag before *supply* responds, cause *supply, demand* and *price* all to oscillate before the system stabilises.

The results shown on the last four pages are all derived from a single simulation model based on the causal loop diagram shown on page 118 – and these are just a few of an enormous number of different possibilities. This is a vivid illustration of how a single simple causal loop diagram can explain many highly complex real behaviours.

Price in Context

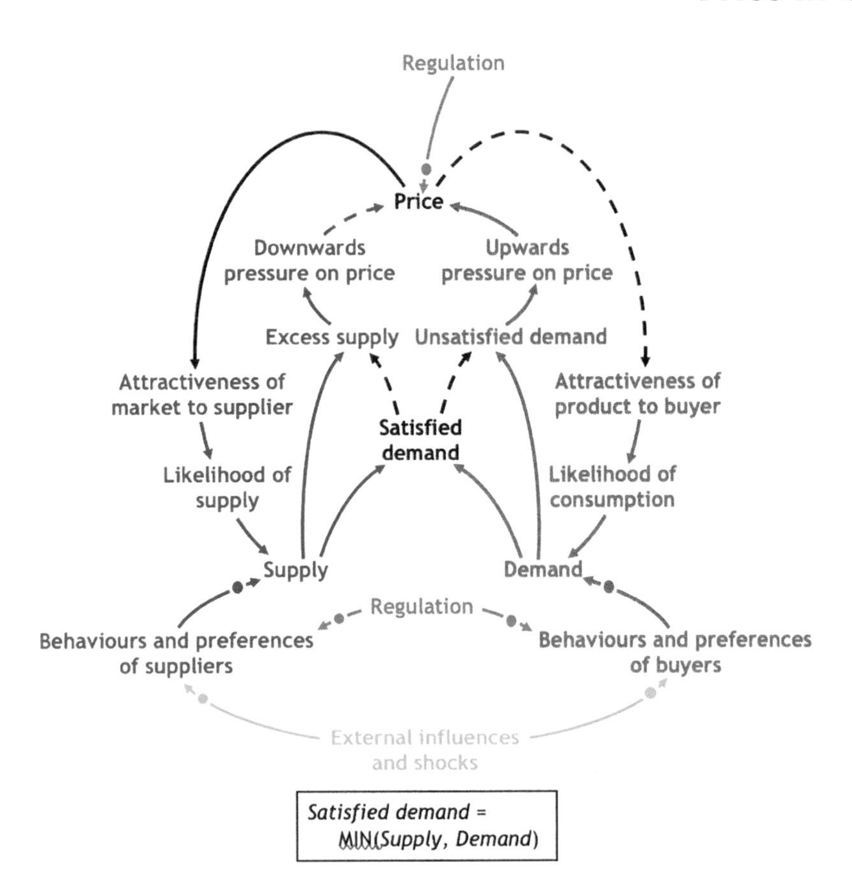

This causal loop diagram shows *price* in a broader context. *Supply* and *demand* are influenced by the *behaviours and preferences of suppliers* and *buyers*, respectively, and each of these react to *external influences and shocks*.

Furthermore, *regulation* can directly influence *price* (for example, taxes and *price* controls), *supplier behaviour* (banning cartels…) and *buyer behaviour* (rationing, controlled substances, planning laws…).

Also, *price* does not just refer, for example, to the *price* of commodities and goods such as oil: wages are *prices* too, in the market for labour, as are commercial interest rates (*demand* being the need for capital for investment, *supply*, the provision of capital), and currency exchange rates (the *demand* being the requirement for [this currency], and *supply*, the willingness of those who hold [this currency] to exchange it for the currency being offered).

Note that all the newly introduced links are influence links.

Monopolies and Monopsonies

As vividly illustrated by the stock market, the behaviour of *prices* over time can be very complex. But underpinning this complexity is the causal loop diagram shown on page 126, from which a key inference is that *price* does not depend directly on either *supply* or *demand*: rather, the instantaneous *pressure to change the price* depends on the difference between *supply* and *demand*, such that a positive difference, *excess supply*, drives a *downwards pressure on price*, while a negative difference, *unsatisfied demand*, drives an *upwards pressure on price*. These are mutually exclusive, in that there is either *excess supply* or *unsatisfied demand*: both cannot exist simultaneously. Accordingly, at any time, there is either a *downwards pressure on price* or an *upwards pressure on price*. And, given delays as regards, for example, the time needed to adjust *supply* capacity, or to run down stocks, the 'signals' transmitted through the system might not be well synchronised, with the result that the *price* might oscillate gently as shown on page 125, or yo-yo more violently. It is only when there is neither *excess supply* nor *unsatisfied demand* – as happens when *supply* and *demand* are equal over some period of time – that the *price* is stable.

Technically, the system illustrated on page 126 comprises two pairs of balancing loops – one pair being invoked by *excess supply*, and the other by *excess demand*. At any time, either one or the other is in operation, or all are 'switched off', as happens when *supply* and *demand* are in balance.

Monopolist suppliers can exploit *unsatisfied demand* by maintaining *supply* just below *demand*, benefiting from a continuous *upwards pressure on price*. This can be countered by a *regulator*, legally preventing single monopolists or cartels, in which otherwise independent organisations agree on how much *supply* each will contribute to the market, and often on *prices* too. Alternatively, within a competitive market, individual enterprises will attempt to differentiate their offer to create a 'local monopoly' that escapes the regulator's scrutiny – hence the power of brands. And a special case of a monopoly supplier is a country's national bank, the sole source of money, and the setter of its *price*, the base interest rate.

Monopsonist buyers – notably governments – can exploit the market the other way. If two or more suppliers bid for a single contract, there is *excess supply*, with the *downwards pressure on price* benefiting the buyer. But once the contract has been awarded, the balance of power changes, and the chosen supplier is now, in effect, a monopolist. So as soon as something happens that was not within the original specification, something that the buyer really needs, there is now a local *unsatisfied demand*, and the contractor is in a very strong position to charge a high price accordingly – which is one of the reasons why so many large public sector contracts end up costing far more than was originally expected.

Ultimately, much can be explained by the single, simple causal loop diagram shown on page 126.

A Simpler Representation

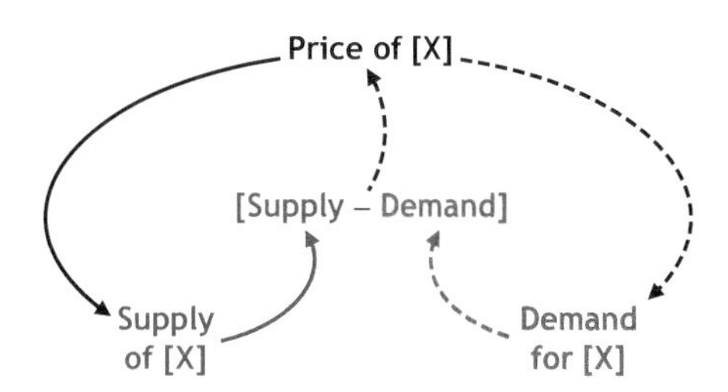

This causal loop diagram is a briefer version of the *supply, demand, price* causal loop diagram shown on pages 118 and 126.

The 'interior' has been simplified to a single variable, the difference between *supply* and *demand*, written as [*Supply – Demand*].

If this is zero, *supply, demand* and *price* are stable; if positive, the excess supply results in a reduction in *price*; if negative, the unsatisfied demand results in an increase in *price*.

The 'outer' links have been simplified too, with an increase in *price* stimulating an increase in supply and a decrease in *demand* (perhaps with delays); likewise, a decrease in *price* results in a decrease in *supply* and an increase in *demand*.

As expected, the structure of this system is a pair of interconnected balancing loops, both acting simultaneously.

Supplier Costs

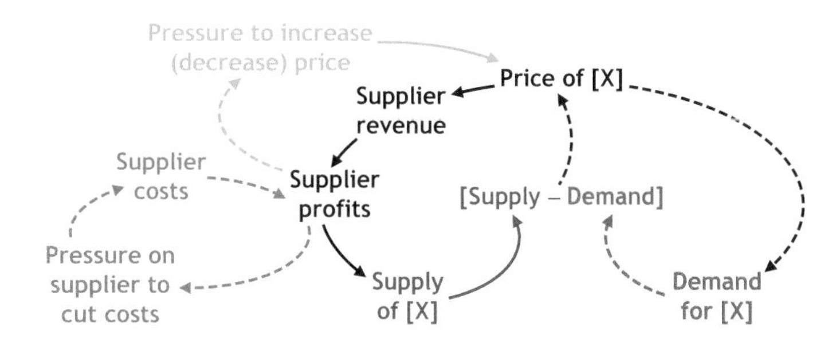

As noted on page 119, the causal loop diagrams relating to *price* discussed so far have assumed that *supplier costs* are constant; this diagram now takes these *costs* into consideration.

For any supplier, the higher the *price*, the greater the *revenue*, from which *profits* are determined by deducting the corresponding *costs*. If, however, there is a fall in *price*, or if *costs* rise (for example, owing to an increase in the price of a relevant resource, such as a component, a raw material, energy or labour), then *profits* fall.

In the short term, this reduction in *profits* might just be absorbed. Most suppliers, however, will seek to restore their *profits*, and one way of doing this is to *cut costs*, as shown by the balancing loop shown in magenta to the lower left. This loop acts to maintain *profits*, without affecting *supply*, *demand* or *price*, and so there is no impact on the larger economic system.

Alternatively, the supplier may seek to recover any fall in *profits* by *increasing the price*. This invokes a second balancing loop, but also triggers all the 'market-facing' loops affecting *supply*, *demand* and *price* as discussed on the previous several pages. And should *profits* rise (attributable, say, to a fall in *costs* resulting from a fall in the price of a resource, an increase in internal efficiency, or innovation), the supplier might choose to invest the 'surplus' as a *decrease in price*, perhaps with the intention of gaining a competitive advantage as discussed on pages 84 and 85.

Wages and Prices

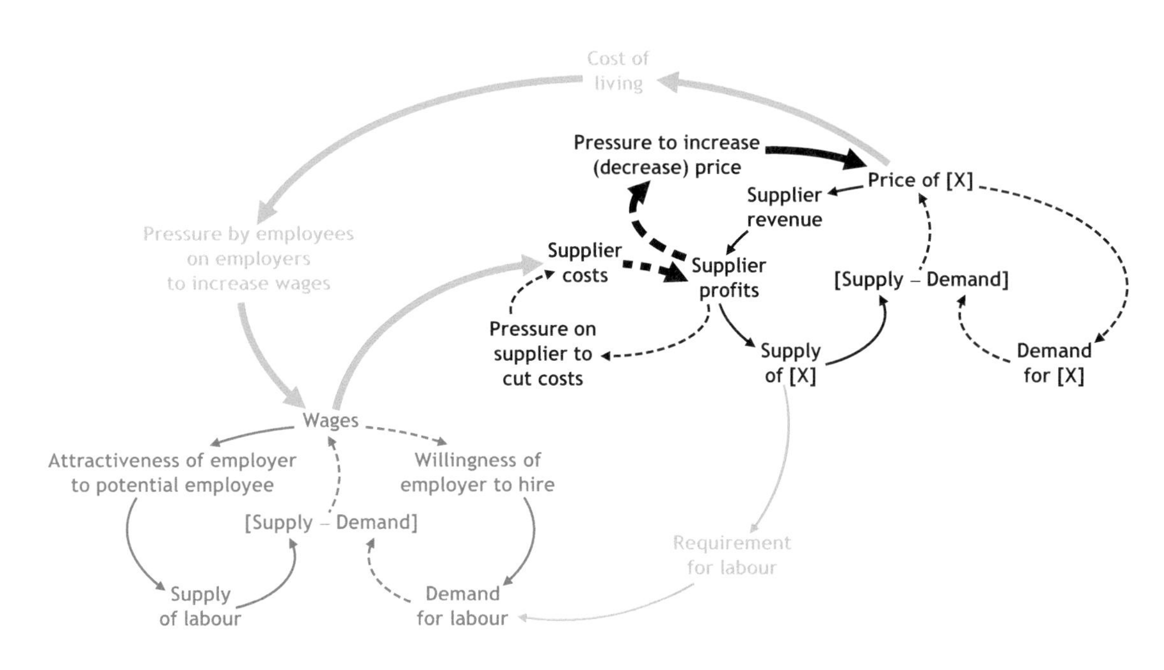

The highlighted loop can drive *wages* and *prices* up, but without affecting either *supply* or *demand*.

For narrative, see following page

Wage-Price Inflation

Labour is an important *cost* to every organisation, and labour too has a *supply* (those people who seek work), a *demand* (employers who need workers), and a corresponding *price* (the level of *wages*). The diagram on page 130 therefore shows how the market for labour (magenta, lower left) interacts with the market for the product or service (black, upper right) for which that labour is needed.

In general, the higher the *wage* level, the *more attractive an employer offering that wage will be to a potential employee*, and so the greater the potential *supply of labour* to which that employer will have access. At the same time, a higher *wage* level is likely to reduce the *willingness of an employer to hire new staff*, whilst the greater the *willingness*, the greater that employer's *demand for labour*.

To the upper right, in black, is the diagram shown on page 129, representing the perspective of a manufacturer of some product [X], or the deliverer of a service.

As can be seen, there are three important connections, shown in green, between the 'black' and 'magenta' systems.

First, for a single supplier, whatever product is manufactured, or whatever service is delivered, there will be a requirement for the corresponding resources, one of which will be labour. The quantity of product or service *supplied* therefore results in a *requirement for labour*, so determining the *demand for labour* – as shown in the lower central part of the diagram.

Second, the *wages* paid to the employees contribute to the *supplier's costs*, so the higher the *wages*, the greater the *costs* and the lower the *profits*. This puts *upwards pressure on price*.

The diagram also shows a 'big picture' effect, relating to the whole economy: if the right-hand side represents the productive economy in its entirety, and the left the whole labour force, then *price* represents the overall *price* level, and *wages* the overall level of earned income. The higher the overall *price* level, the higher the overall *cost of living*, one consequence of which is for those in work to *demand higher wages*, perhaps by mutual agreement, perhaps by threatening to withhold their labour by striking.

As can be seen, there is a closed loop from *price* through *cost of living*, *pressure to increase wages*, *wages*, *supplier costs*, *supplier profits*, *upwards pressure on price* and back to *price*. This closed loop has two inverse links and so is a reinforcing loop, one behaviour of which is to drive the growth of both *wages* and *prices* – growth which does not necessarily have any impact on overall *supply*, or the economy's aggregate *demand*. This is wage-price inflation.

Economic Depression

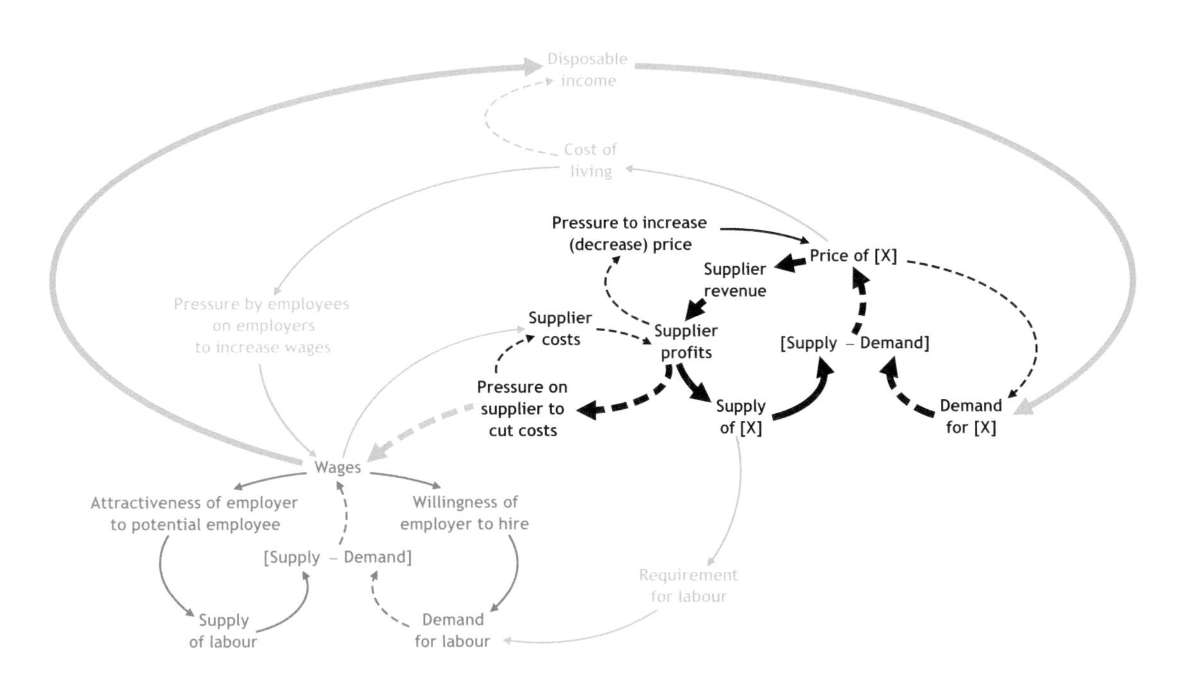

This network of loops can drive *wages, prices, supply* and *demand* down together, so depressing the entire economy.

For narrative, see following page

Depression – Or Growth?

The causal loop diagram on page 132 adds just one feature to the diagram on page 130, the economy's aggregate *disposable income*. That's important because the lower the overall *wage level*, the lower the *disposable income* within the economy, and the lower the aggregate *demand*.

This, in turn, makes the difference [*Supply – Demand*] increasingly negative, so reducing *prices*. *Revenues* and *profits* decline, increasing the *pressure to cut costs*, which can result in a reduction in *wages*. This completes a closed feedback loop with four inverse links: from *demand* to [*Supply – Demand*]; from [*Supply – Demand*] to *price*; from *supplier profits* to *pressure to cut costs*; and from *pressure to cut costs* to *wages*. This loop is therefore a reinforcing loop, one behaviour of which is to drive a fall in *wages, demand* and *prices* – a loop whose impact is somewhat tempered by the action of the associated balancing loop from *price*, through *cost of living*, *disposable income*, *demand*, [*Supply – Demand*] and back to *price*: a fall in the *cost of living* makes money go further.

But that's not all. The fall in *demand for [X]*, and the resulting fall in *price*, triggers the supply-side loop whereby a reduction in *supplier profits* drives a reduction in overall *supply* too, as highlighted on the centre-right.

In this system, *supply*, *demand*, *prices* and *wages* all fall together – this being what the economists call a 'depression', which, once triggered, can be very hard to arrest, as witnessed by the history of the late 1920s and early 1930s, and more recent history too.

Reinforcing loops, however, don't necessarily decline. They can grow too. Suppose, then, that 'something happens' to cause *wages* to rise. This increases *disposable income*, driving up *demand,* ultimately resulting a *rise* in *prices*. *Supplier profits* increase, and the *pressure to cut costs* becomes a negative number, so exerting a negative *downwards pressure on wages*, or, in more normal language, raising *wages* – so increasing both *disposable income* and *demand*. And simultaneously, the supply-side loop gets going too, as *supply* rises to meet the increasing *demand*. In this scenario, *supply*, *demand*, *prices* and *wages* all rise together, this being the economic growth that every government wishes to deliver to its citizens.

A single diagram, that shown on page 132, represents both economic growth and depression, just as the (very much simpler) diagram on page 44 represents both the boom and bust of a single business – vivid examples of the power of systems thinking in taming the complexity of real systems.

The real economy, of course, is vastly more complex than as depicted on page 132 – (un)employment and taxation, for example, are just two of many important factors that are not shown. But I trust this diagram is a useful start, a diagram that you can enrich…

Interest Rates, Exchange Rates and Currencies

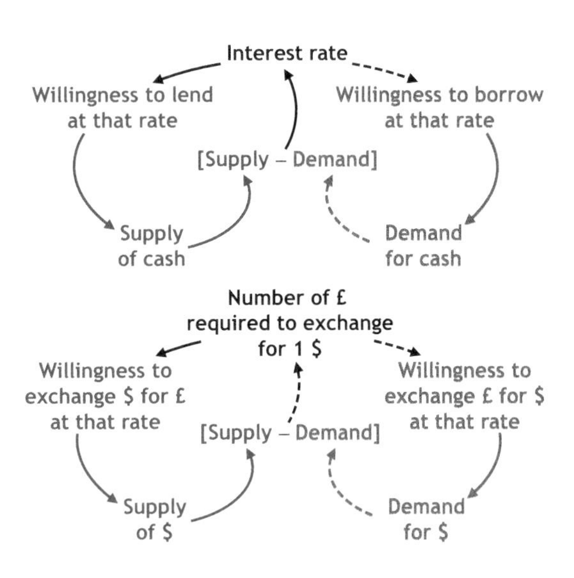

The *supply, demand, price* diagram on the upper left corresponds to the *lending* and *borrowing of cash,* with the price being the *interest rate*; below, a diagram for *currency exchange rates* for the sale of US$ for £ sterling. In these two cases, as in all the other examples in this chapter, the seller, or supplier, is in blue on the left (in these cases the lender of cash, and the seller of US$ in exchange for £ sterling), and the buyer is in red on the right (the borrower, and the person who is wishes to exchange £ sterling for US$).

So far, all the *supply, demand, price* diagrams have been structurally similar, and this diagram applies to all traded commodities, for which the appropriate *price* is designated as so-many units of a currency. But what is the diagram for a currency itself? And what is the corresponding price?

My suggestion is shown in the causal loop diagram below, in which the *price* is the *perceived value of each unit of currency,* representing the belief, shared across users of that currency, that the currency has a particular value, and will maintain it.

Accordingly, the higher that *value*, the smaller the *number of units of currency needed to buy, say a loaf of bread*, and, correspondingly, the smaller the number of units of currency that need to be in circulation, this being the *demand*. Furthermore, for a given level of *demand* for the currency, the scarcer the *supply*, the higher the *perceived value* – which explains why gold, as a currency, is highly valued.

For printed or electronic money, however, and unlike all other traded items, the quantity of money in circulation, the *supply*, is determined not by a price, but by *government policy*. And sometimes, as discussed on the following two pages, that can be problematic…

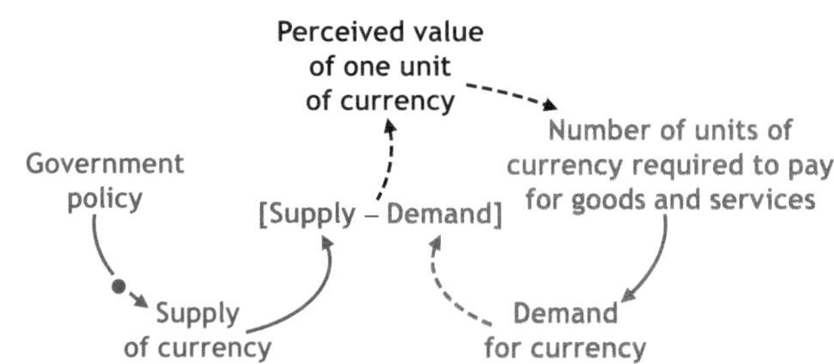

German Hyperinflation, 1918–23

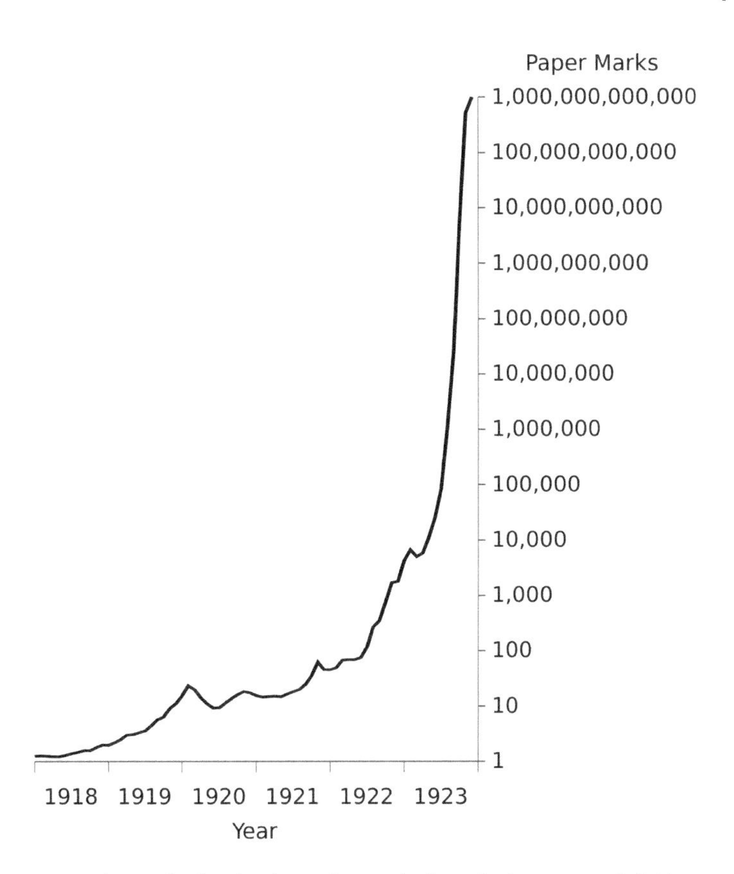

Source: https://upload.wikimedia.org/wikipedia/commons/0/05/Germany_Hyperinflation.png.

In the early 1920s, Germany suffered catastrophic hyperinflation of its paper currency.

This is vividly illustrated by the chart to the left, the (logarithmic) vertical axis of which shows the number of paper marks required to purchase one gold mark.

Towards the end of 1918, about two paper marks would buy one gold mark. By the end of 1923, that same one gold mark would require 10^{12} paper marks.

If the underlying system were a single reinforcing loop with a constant growth rate, a logarithmic plot would be an upward-sloping straight line.

But this graph is not a straight line – the logarithmic plot itself slopes progressively more steeply upwards.

This suggests a structure of at least two linked reinforcing loops, with the growth rate of one being embedded in the other, so that this growth rate itself grows exponentially.

Currencies and Hyperinflation

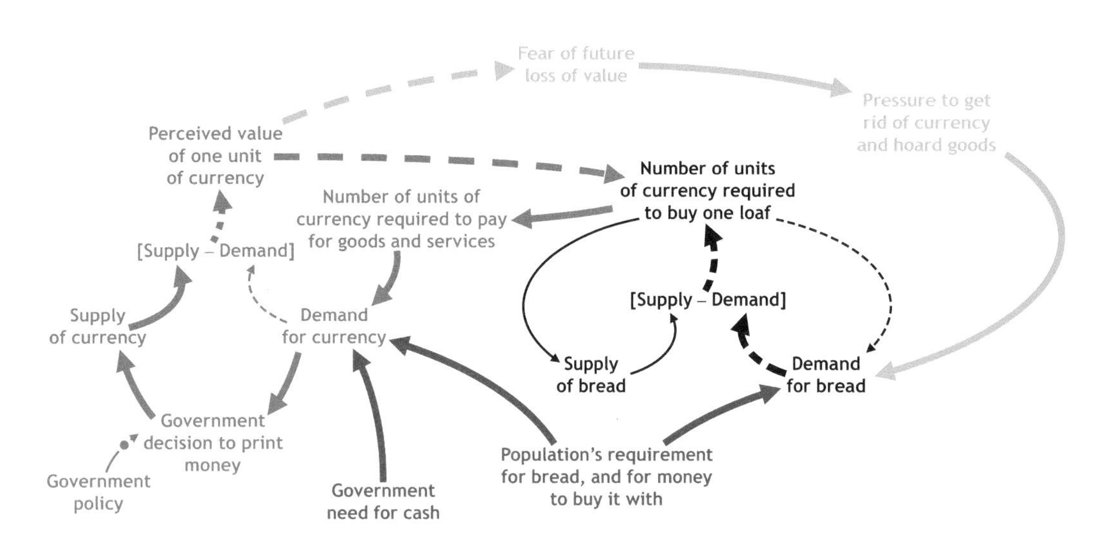

This diagram shows two reinforcing loops, operating simultaneously in the same direction, and amplifying each other. The first is in magenta, from perceived value to units of currency to buy one loaf, and back to perceived value. The other is highlighted first in green and black, and then in magenta, from perceived value to fear of future loss of value, to demand for bread and units of currency to buy one loaf, then through government decision to print money back to perceived value.

The clue, given by the chart on page 135, that hyperinflation is driven by (at least) two reinforcing loops, is validated by this causal loop diagram, which captures the essence of hyperinflation.

The structure shown in black at the centre-right is the familiar *supply, demand, price* diagram for a representative traded item, a loaf of bread, for which the *price* has been more explicitly represented as the *number of units of currency required to buy one loaf.* To the left, in magenta, is a version of the *supply, demand, price* diagram for the corresponding currency, based on that shown in the lower right of page 134.

Hyperinflation

With reference to the loop in magenta, the inverse link from *perceived value* to *number of units of currency required to buy one loaf*, captures the likelihood that if the *perceived value of the currency* falls, then a supplier of bread will require a greater *number of units of that currency for each loaf* – as a community loses *trust in its currency*, *prices* rise. Then, the direct link from the *number of units of that currency for each loaf* to the *number of units of currency required to pay for goods and service* represents that the higher the *price of bread*, the greater the number of *units of currency required*. The combination of these two links – one inverse, the other direct – therefore has the same effect as the corresponding link single inverse link in the diagram to the lower right of page 134.

A second difference is the recognition that the total *demand for currency* is not solely determined by the needs of the general population alone, for the *government requires money* too: governments have to fund their own payroll, as well as obligations such as state pension schemes and the servicing of debt. If, say, debt payments need to be made in a different currency, the government has to exchange the local currency for, say, US$. As the diagram on the centre left of page 134 shows, an increase in *demand* for US$ will increase the *number of units of local currency required for each $*, which, from the government's standpoint, is an unfavourable move in the exchange rate. If the sums involved are large, the government's requirement might weaken the exchange rate in this way, increasing the *demand* for the local currency even more, and with adverse effects on the wider economy too.

The overall *demand for currency* – the total of the *domestic* and *government requirements* – needs to be met. Governments, however, have a solution to an otherwise unsatisfied demand, for they own the printing press: they can *decide to increase the supply of currency* to meet the required *demand* by printing it. But if they do so unwisely, this can be catastrophic…

…for as the diagram shows, the two links from *demand for currency*, through *government decision to print money*, to *supply of currency* complete a reinforcing loop in which an increase in the *supply of currency* causes a reduction in its *perceived value*, driving an increase *in demand*, and the *printing of even more money*. That's not all: at the same time, a decrease in *perceived value* increases *fear that the currency will lose even more value in the future*, driving people to *get rid of their currency* by purchasing 'real' goods. This increases the *demand* for those goods, pushing *prices* up yet further, so completing a second reinforcing loop. As anticipated on page 135, these linked reinforcing loops can grow 'super exponentially' very, very quickly, for together, these two loops drive hyperinflation – as experienced, for example, by Germany in 1922/3, as illustrated on page 135, by Zimbabwe in 2007/8 and by Venezuela in 2016/19. And once these loops start spinning, they can be very, very hard to arrest.

Chapter 11

Conflict – and Teamwork

DOI: 10.4324/9781003304050-13

The Story of 'Red' and 'Blue'

A yell.

Chris turned down the heat under the cooking pan, and rushed into the living room.

Sandy was close to tears; Pat looked sullen.

'I don't want to watch this stupid film!' wailed Sandy. 'I want a cartoon! Pat won't let me have the remote!'

Chris took the remote from Pat and switched to the cartoon.

'Right', Chris said. 'We'll watch the cartoon for the next half-hour, then go to the film. Is that OK?'

Sandy and Pat nodded, and Chris went back to the cooking.

Ten minutes later, another yell.

'Pat snatched the remote back!!!'

The following pages tell the story of '*Red*' and '*Blue*'.

'*Red*' and '*Blue*' each has their own goals, objectives and targets, and the story tells what might happen as time evolves.

As the story unfolds, you might like to think who '*Red*' and '*Blue*' might be.

And you will see that the story has seven different possible overall outcomes.

Which do you prefer?

Red Has an Obligation to Meet a Target...

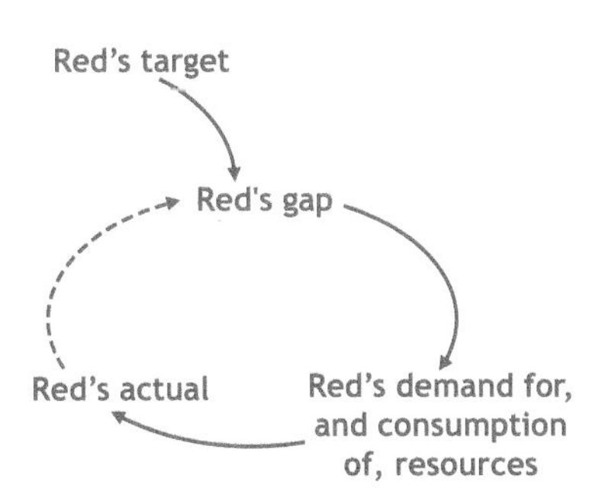

This story starts with *Red*, who wishes to meet some form of *target, objective* or *goal.*

To close the *gap*, and so meet the *target, Red needs resources...*

...and whenever there is a *target*, there is a corresponding *actual*, and – in general – a *gap* between them that needs to be closed.

This represents a balancing loop, designed to bring the *actual* into line with the *target.*

Note that throughout this chapter, the term '*resources*' is being used very broadly, to represent *any resource* required by the organisation to meet its objectives.

This therefore includes:

- staff
- intellectual property
- physical plant and infrastructure assets
- products and services to deliver
- customers and markets to deliver them to
- access to those customers and markets
- funds for investment
- access to influencers or decision-makers
- ...

...and so Does Blue...

The situation for *Blue* is identical, for *Blue* has a *target* too, and needs *resources* to meet them. And as time passes, both *Red* and *Blue* meet their targets and, as they say in the story books, 'everyone lives happily ever after'.

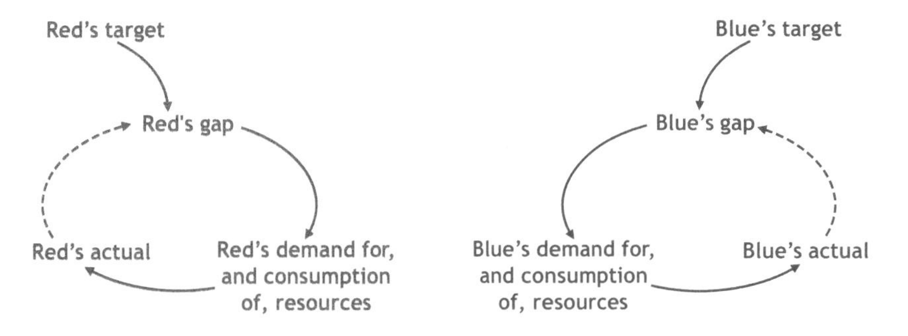

This is the first possible outcome: both Red and Blue successfully meet their respective targets and achieve their respective goals.

So, for example, *Red* might be a manager who meets a manufacturing *target* by ensuring that [so many] high quality products are completed this week, using *resources* of, for example, raw materials, labour, machine capacity, energy and time; meanwhile *Blue* might be a colleague whose *target* is to agree [this value] of orders with a customer, using *resources* of a car (to visit the prospective customer) and time, so that those products can be sold. Or *Red* might be a teacher whose immediate *objective* is to deliver a really good lesson, using *resources* of the classroom and a video, whilst *Blue* might be someone on holiday, whose *objective* for the day is to do as little as possible, consuming only *resources* of some delicious food, some cool drinks, and some space on the beach.

'Happy ever after' is by far the most common outcome, and happens all the time, but it is not the only possible outcome...

...and Have to Compete for Resources...

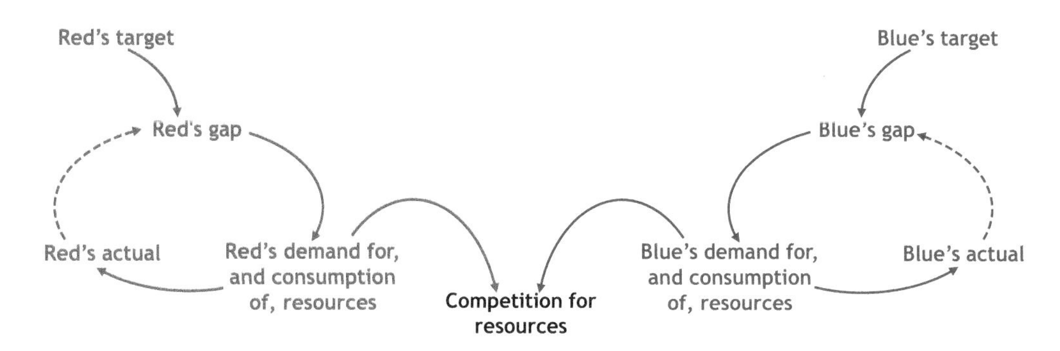

It may happen, however, that in seeking to meet their respective *targets*, *Red* and *Blue compete for resources*, for example, skilled staff, investment funds, or customers...

...Which Are Limited...

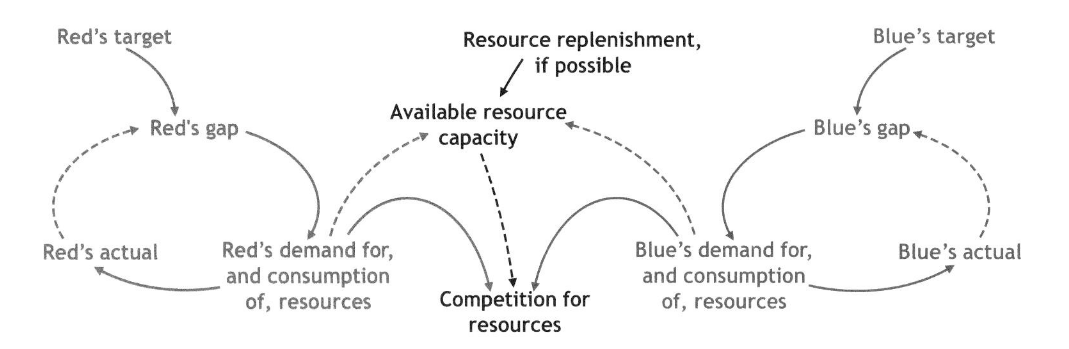

...for, especially in the short term, many *resources* are finite, as measured by the relevant *available resource capacity*. If the *available resource capacity* is significantly greater than the aggregate demand of *Red* and *Blue*, there is enough *resource* to satisfy both requirements, and *Red* and *Blue* can each meet their respective *targets*.

As already noted, in the current context, *resources* can assume many forms. Some are inherently limited, and so any *consumption* must deplete the remaining *available capacity*, as indicated by the inverse links. Examples of inherently limited *resources* include time-bounded *resources* such as *working hours available per person per week* (there are only 24 hours in a day!), and also, for many organisations, *funds available for investment*.

Other *resources* might be limited in the short term, for example, the *stock of raw materials*, or *total staff capacity*, but can in principle be *replenished* by, say, re-ordering raw materials from a supplier, or by hiring additional staff. Depending on what the *resource* is, *replenishment* might take some time, and so short-term *capacity* constraints are often significant. Some *resources* have in the past been considered to be unbounded, for example *water*, which is *replenished* by rain. This does indeed happen, but not universally, as drought-stricken areas know well.

…so Red Starts to 'Play Games'…

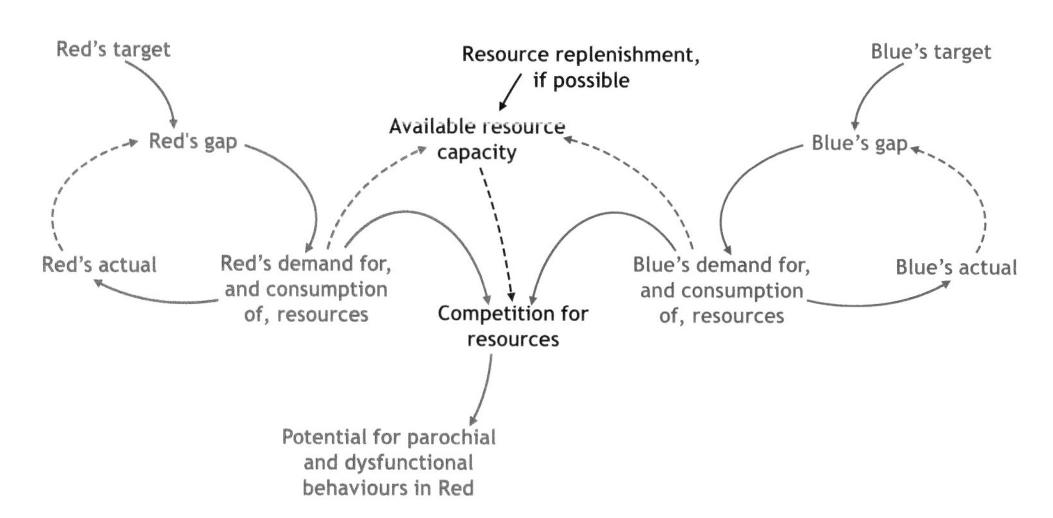

But if *Red* feels that there is not enough *available resource capacity* for its own needs, this can lead to a *potential for parochial and dysfunctional behaviours in Red* with respect to *Blue*: for example, failing to share information, arguing about internal costs and prices, overbidding for investment funds...

...Which Wastes Resources...

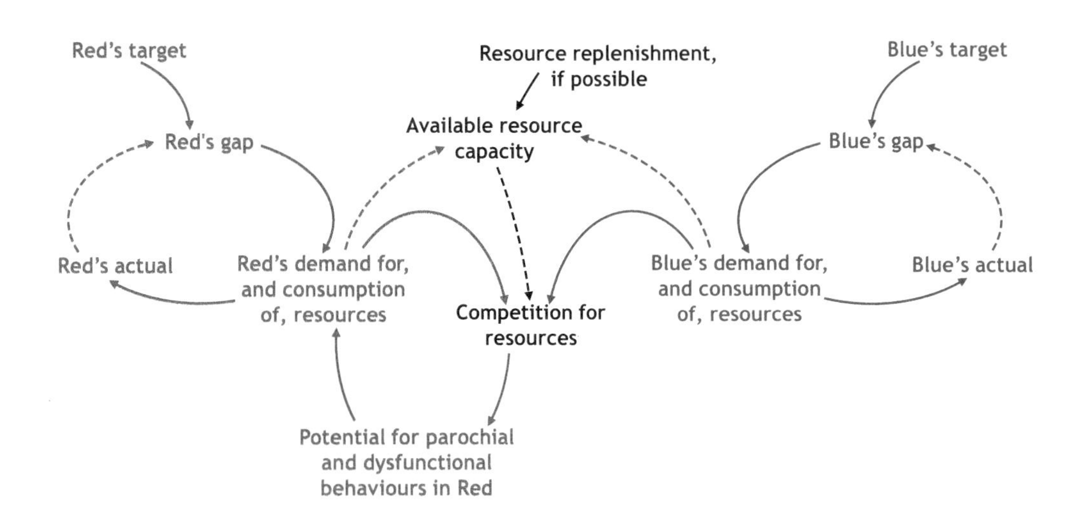

This, in turn, drives *Red's demand for, and consumption of, resources* as *Red* progressively tries to capture as much as possible of the total 'cake'. To take a (very!) trivial example, *Red* might take more pencils than are needed from the stationery cupboard 'just in case we run out...', whilst also denying them to *Blue*...

…and Provokes Blue to Retaliate…

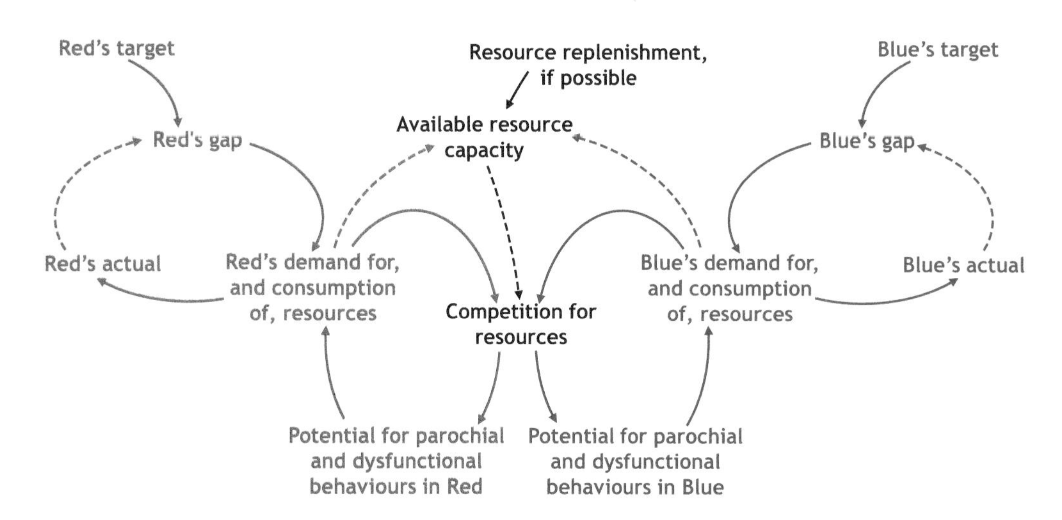

Meanwhile, exactly the same thing is happening for *Blue*...

Two Reinforcing Loops

This creates a system of two simultaneous reinforcing loops, each of which can show either exponential growth or decline with the overall outcome depending on which loops are operating in which mode.

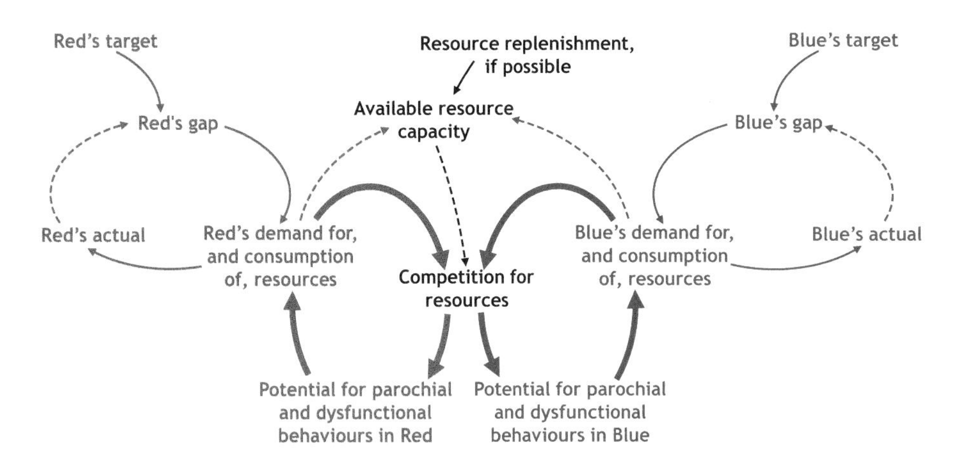

One possibility is that both *Red* and *Blue* consume ever more of the *resource*, with both reinforcing loops, showing exponential growth – as, for example, happens in an 'arms race'...

...until the *available resource capacity* becomes so depleted that very little resource is left, causing both reinforcing loops to flip to exponential decline.

Both *Red* and *Blue* 'lose'.

This is the second possible overall outcome.

Alternatively, *Red* might grab enough *resource* to grow, and deny those *resources* to *Blue. Red's* reinforcing loop shows exponential growth whilst *Blue's* reinforcing loop shows exponential decline...

...until *Blue* is forced out of business, just as the dominant player in many sectors has driven out the 'corner shops'...

Red is the 'winner'; *Blue* is the 'loser'.

This is the third possible overall outcome.

…Which Can Get Nasty…

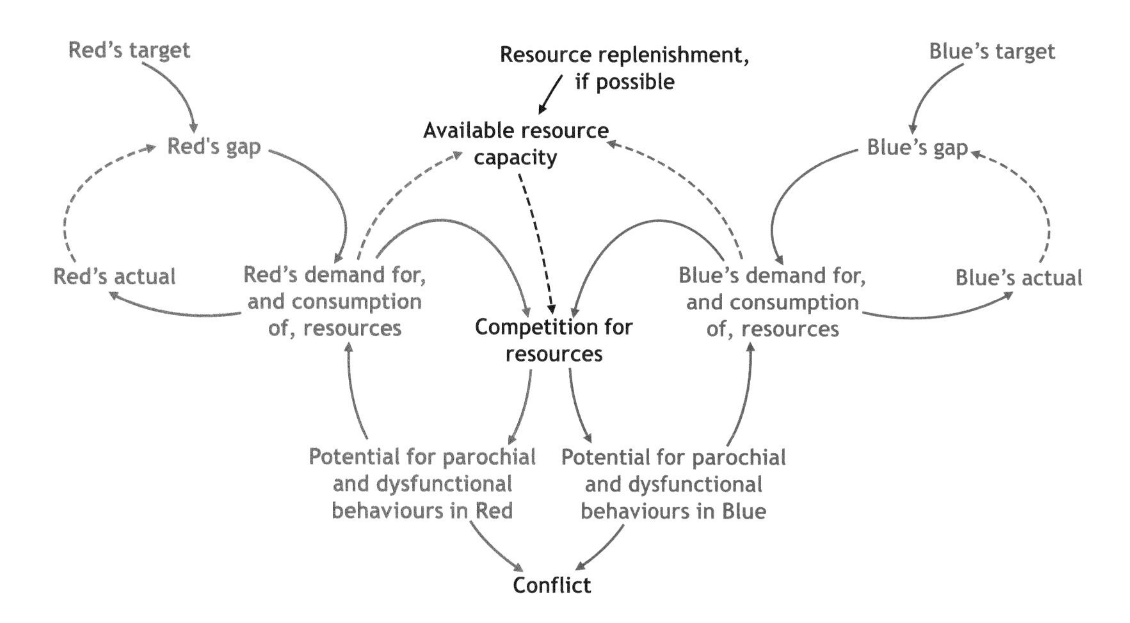

Another possibility is that the mutually *dysfunctional behaviour* escalates into *conflict…*

...Waste Even More Resources...

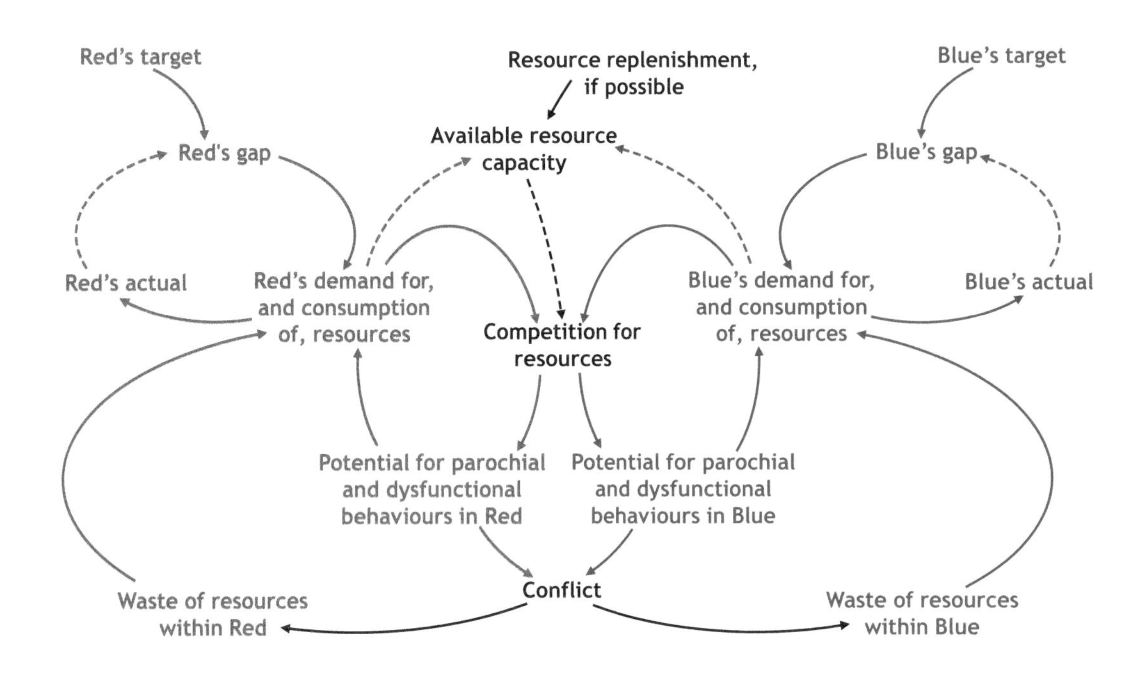

Conflict leads to a *waste of resources* within both *Red* and *Blue*, as, for example, time is wasted on internal arguments, causing both *Red* and *Blue* to be distracted from their true purposes. This, in turn, increases both *Red's* and *Blue's demand for, and consumption of, resources...*

...and Drive Two More Reinforcing Loops...

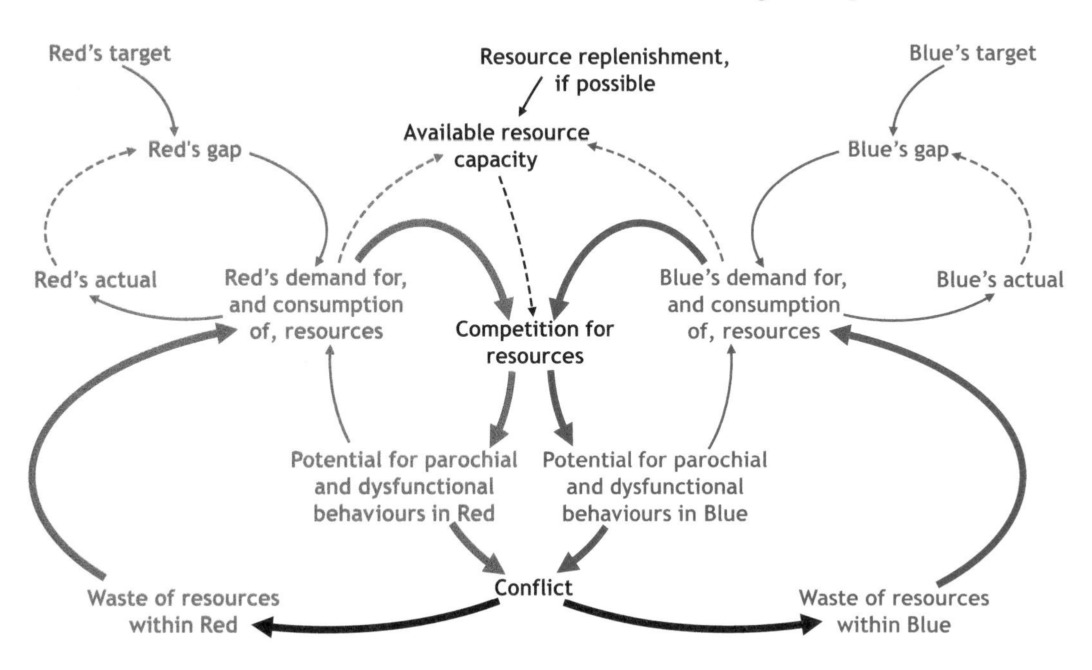

...so forming a second symmetrical pair of reinforcing loops.

Four Reinforcing Loops

Both pairs of reinforcing loops operate together, driven by the two balancing loops which represent the pressure on *Red* and *Blue* to meet their respective *targets*, in competition with one another, so progressively enflaming the *conflict* and squandering the finite *resources*…

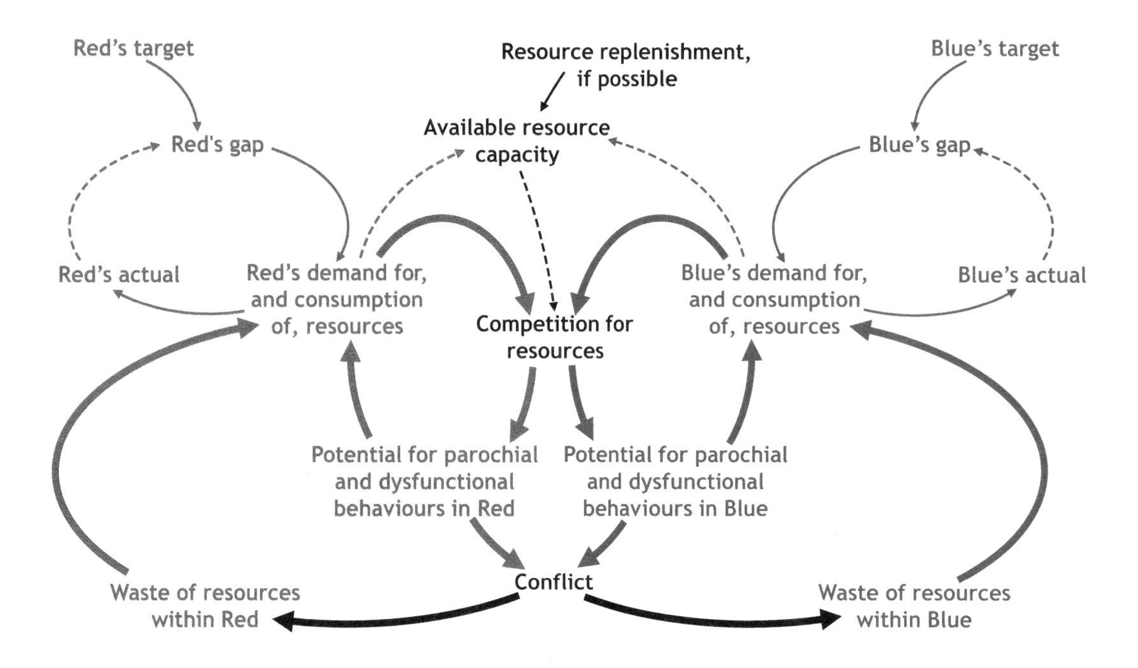

…with the ultimate result of either mutual destruction, in which all the reinforcing loops grow exponentially together, and then together crash into exponential decline…

…or 'winner takes all', in which one set of reinforcing loops grows exponentially whilst the other exponentially declines.

Who Wins? Or Do Both Red and Blue Lose?

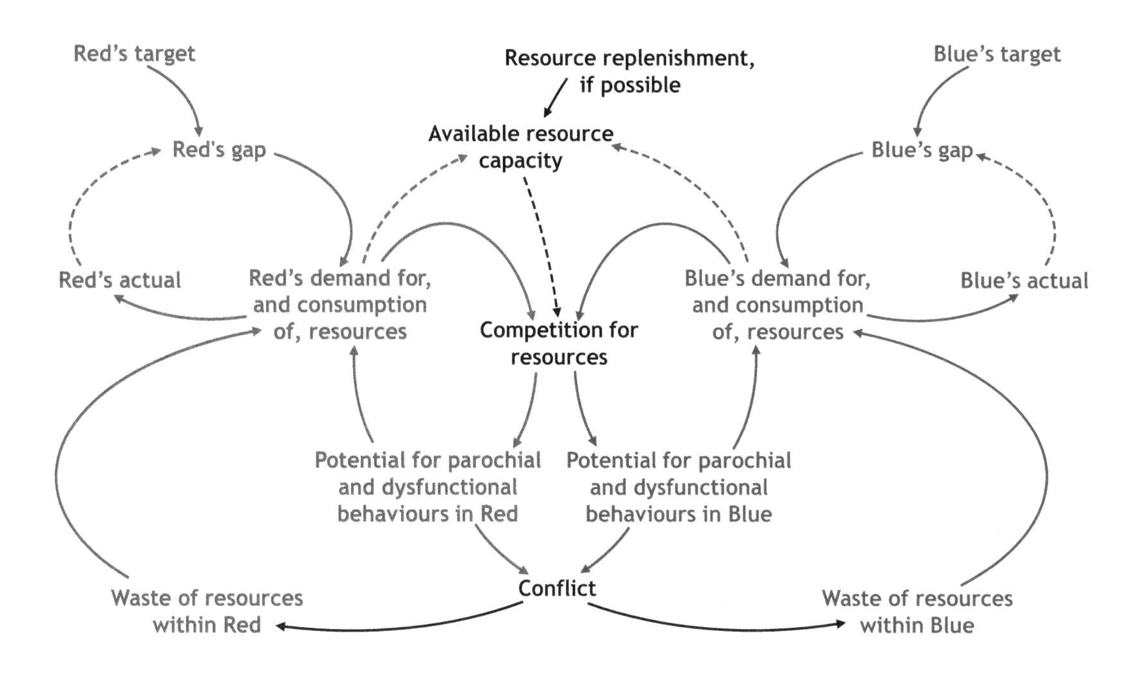

This is the fourth possible overall outcome, and is potentially a very damaging lose-lose game, in which both Red and Blue progressively damage their own, and each other's, ability to survive. Even if there in an eventual 'winner', the 'victory' might have come only at very great cost.

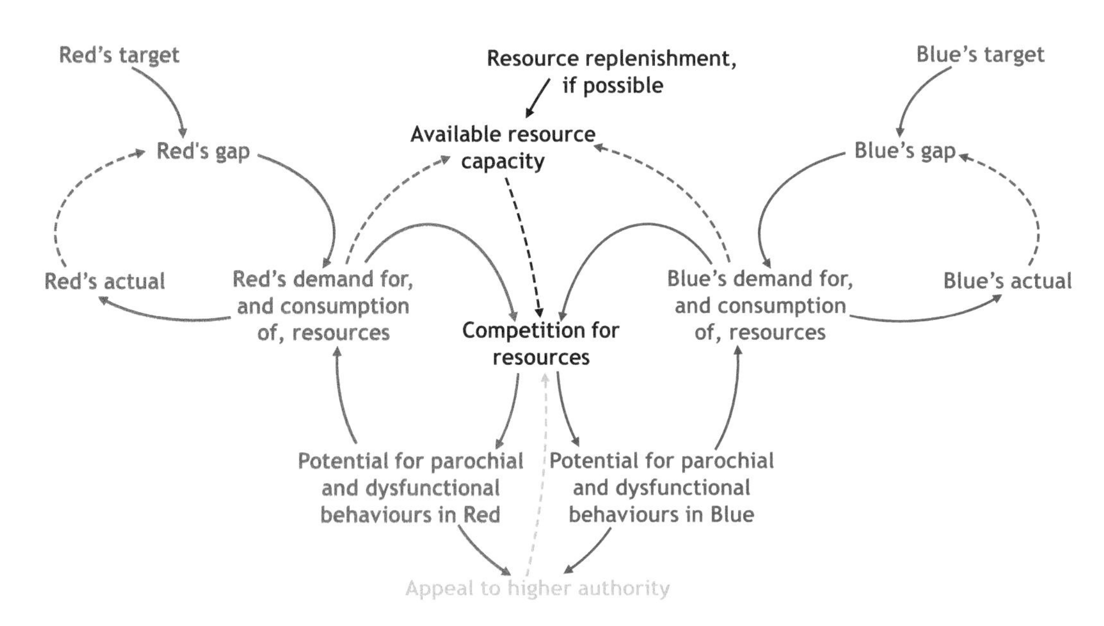

An alternative to *conflict* is an *appeal to higher authority,* the purpose of which is to arbitrate on, and hence reduce, the *competition for resources* by allocating the scarce *resource* between *Red* and *Blue*: hence the inverse link from *appeal to higher authority* to *competition for resources.*

Two Balancing Loops

This introduces a pair of symmetrical balancing loops, which control the action of the two associated reinforcing loops, so stabilising the system.

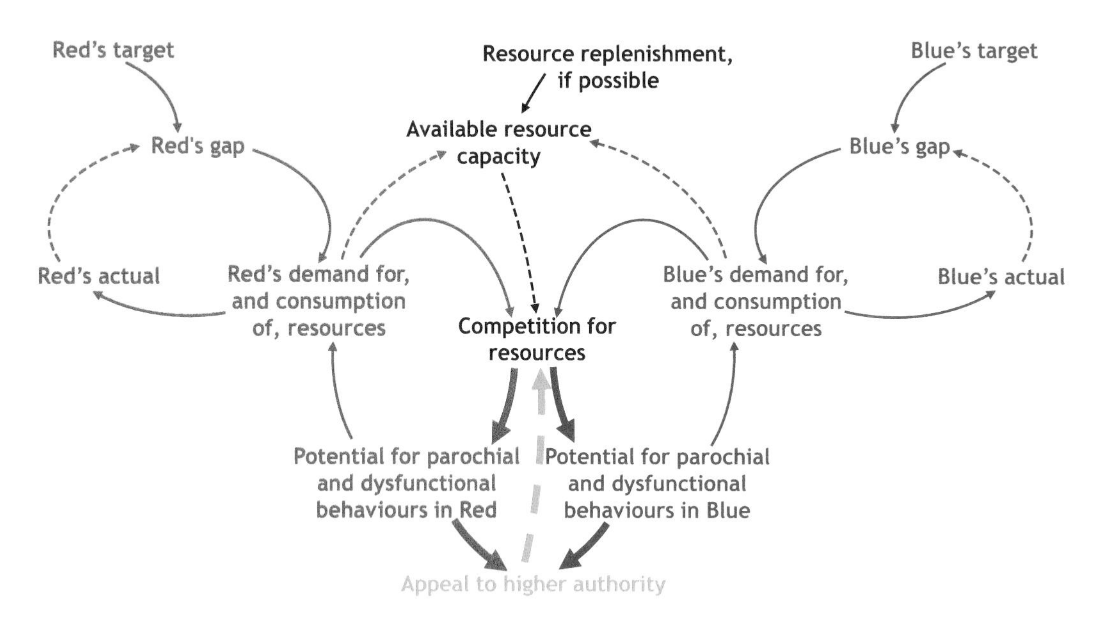

This is the fifth possible overall outcome.

Maintaining this stability, however, usually requires the deployment of some form of 'police force' to ensure that the allocation of *resources*, as determined by the *higher authority*, is not transgressed – as the story on page 140 illustrates. Rather than resolving the *conflict*, this can often act to hold it just under the surface, from which it will erupt once more, given an appropriate opportunity – as happened in the story on page 140 after the 'higher authority' had left the living room. And when those UN troops in blue helmets lose control...

Like conflict, this can be a mess. Surely there is a better way...

…and an Even Better Way of Avoiding Conflict…

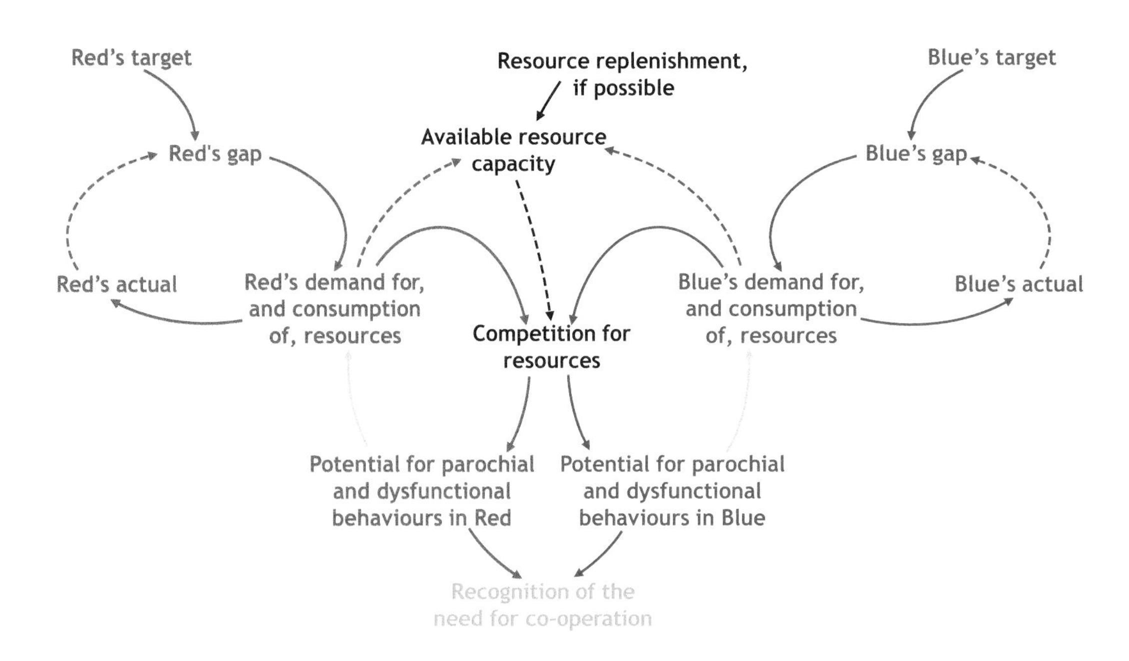

Yes, there is a better way, as illustrated on this page. This is very similar to the diagram shown on page 149, but instead of giving rise to *conflict*, the spectre of the *potential for parochial and dysfunctional behaviours* in *Red* and *Blue* creates a mutual *recognition of the need for co-operation*. This at once significantly weakens the feedback to *Red's* and *Blue's demand for, and consumption of, resources*, this weakening being represented by the pale grey links.

…Is by Co-Operating, Not by Competing…

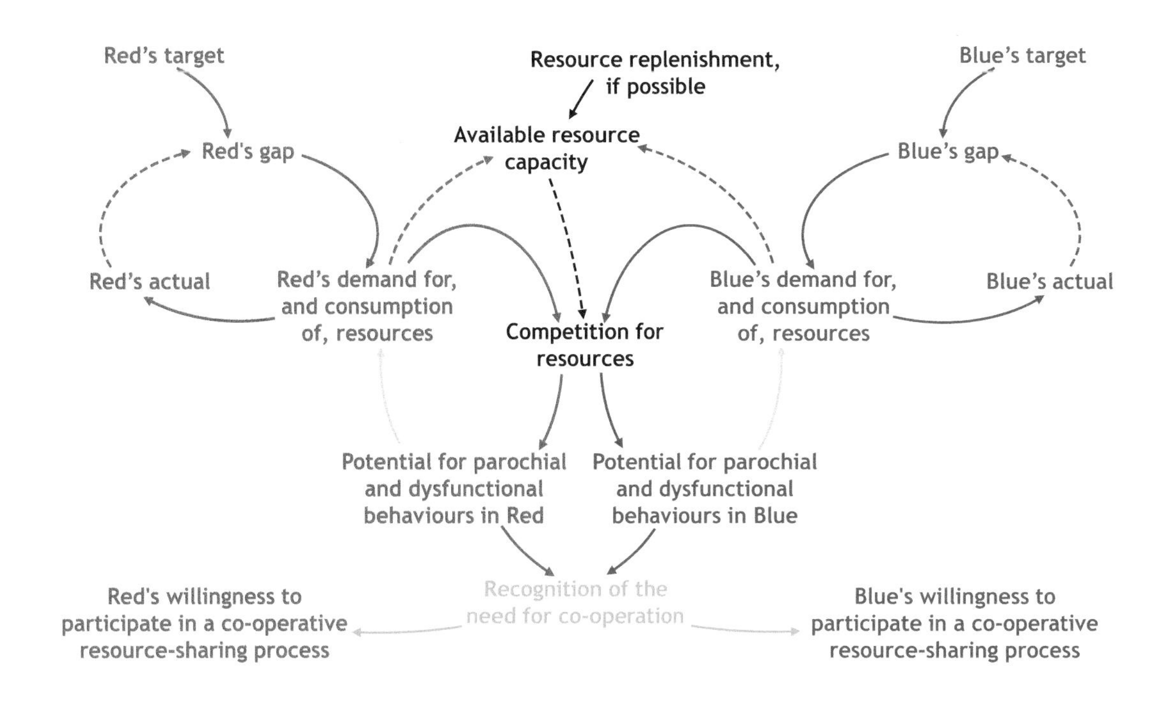

The mutual recognition of the need for co-operation leads to a mutual willingness to participate in a co-operative resource-sharing process, with the shared objective of optimising the available resource capacity…

...for Example, by Agreeing How to Share the Scarce Resource...

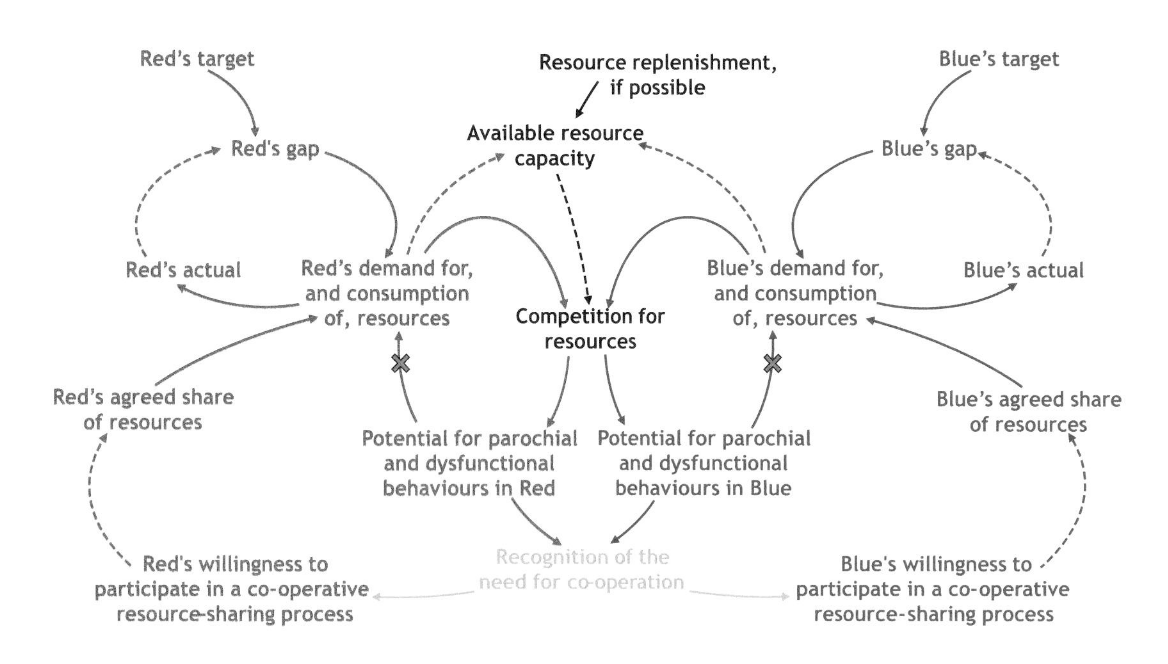

This moderates *Red's* and *Blue's* respective claims, and also serves to 'switch off' the two reinforcing loops from *Red's* and *Blue's potential for dysfunctional behaviour* to the corresponding *demand for resources*...

Two Stabilising Balancing Loops

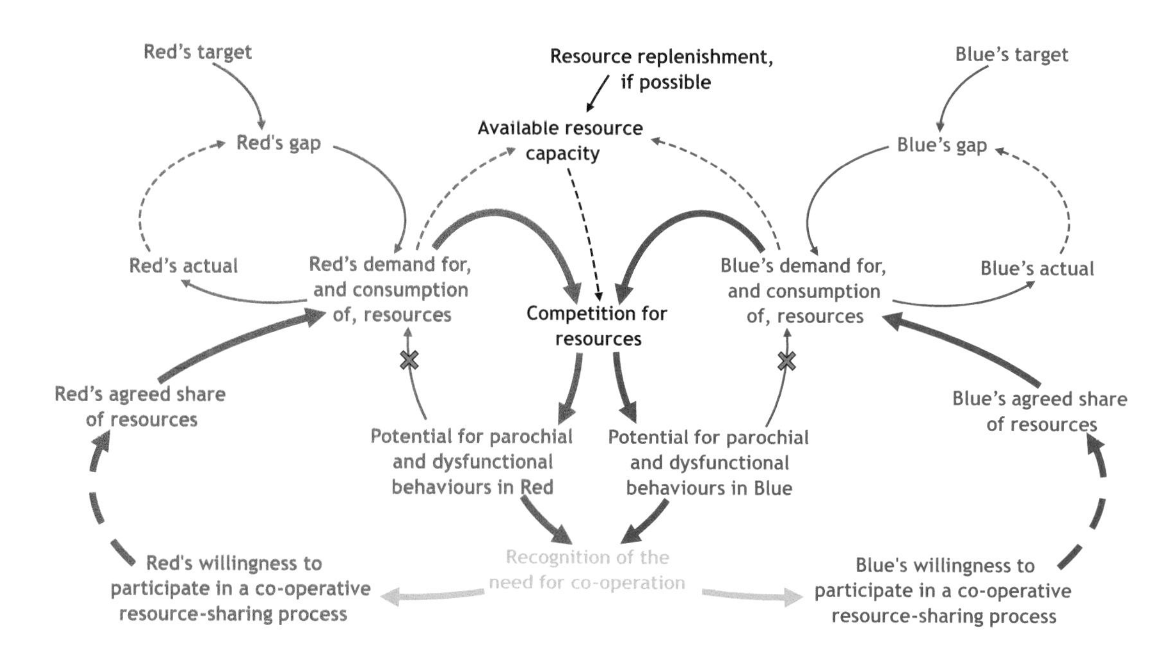

...so creating a system comprised of two symmetrical balancing loops, which stabilise on the agreed shares of the *available resource capacity*.

This is the sixth possible overall outcome.

And Even Better Still...

Alternatively, rather than agreeing on how best to share the limited resources, *Red* and *Blue* might agree that it is more effective to participate in a *co-operative target-setting process*...

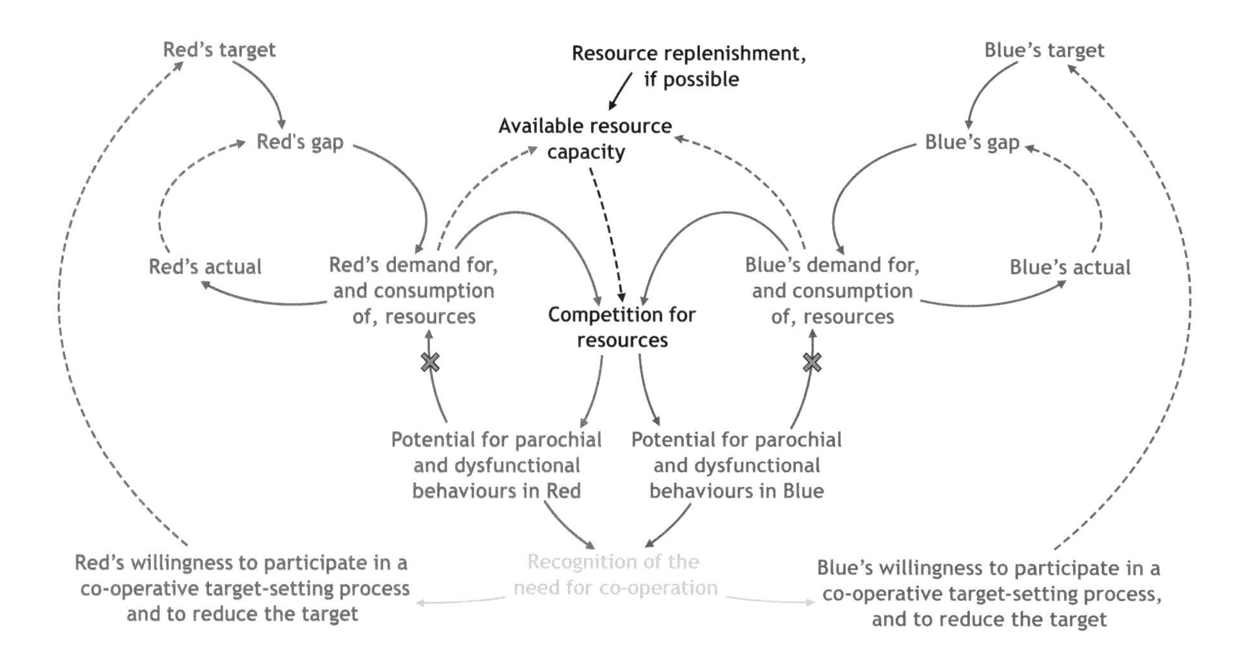

...in which the *targets* that *Red* and *Blue* are striving to meet are mutually agreed, in the light of the *available resource capacity.*

…Is to 'Combine Forces'…

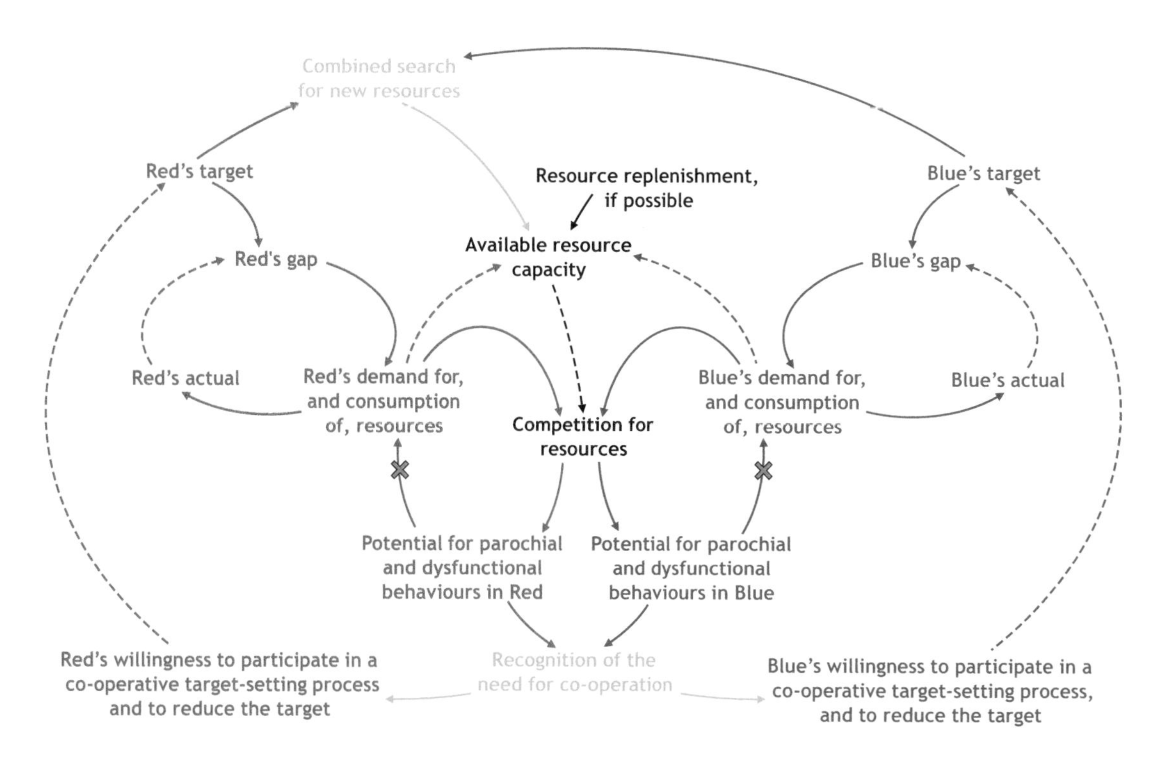

Even better is the possibility that the *co-operative goal setting process* will result in a joint agreement to allocate some of the scarce resources (such as skilled staff or investment funds) to a *combined search for new resources*…

Two Reinforcing Loops

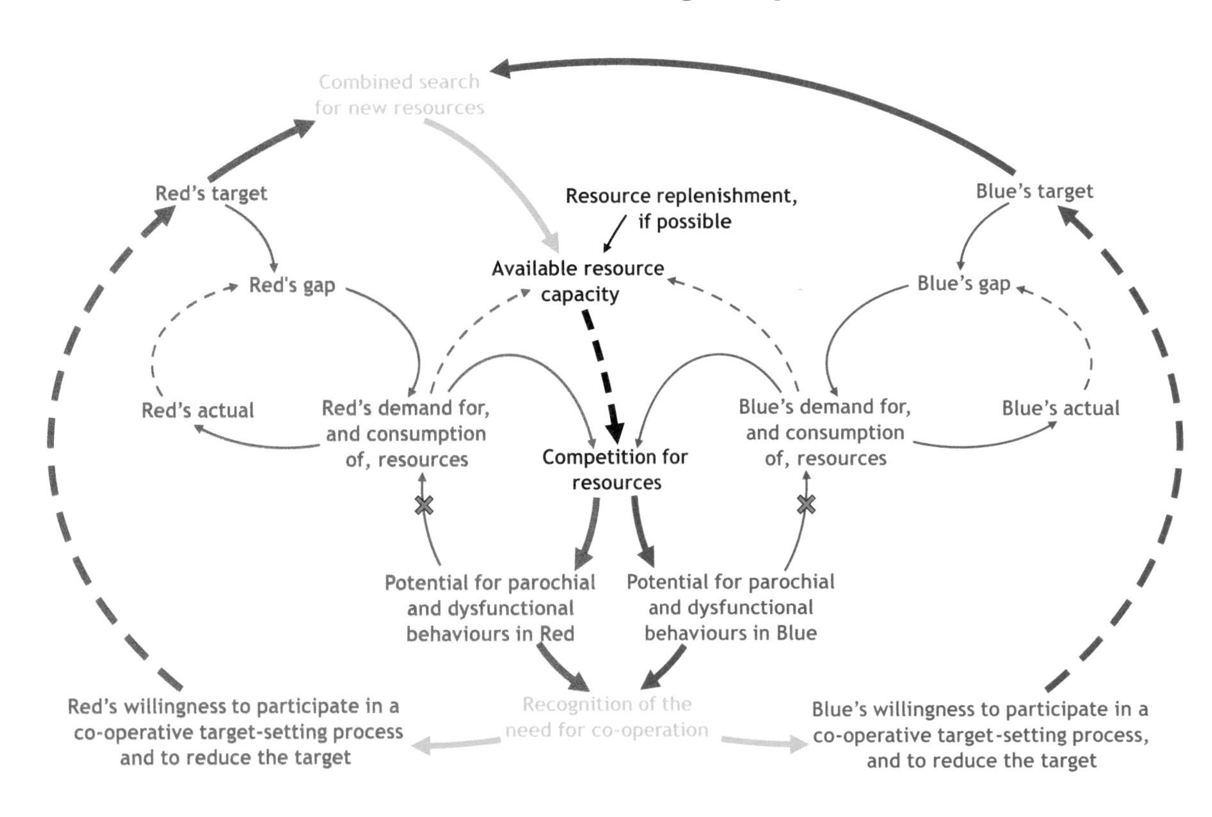

...for this drives two symmetrical reinforcing loops which increase the *available resource capacity*, so alleviating the ultimate constraint to the whole system. *This is the true vehicle of growth.*

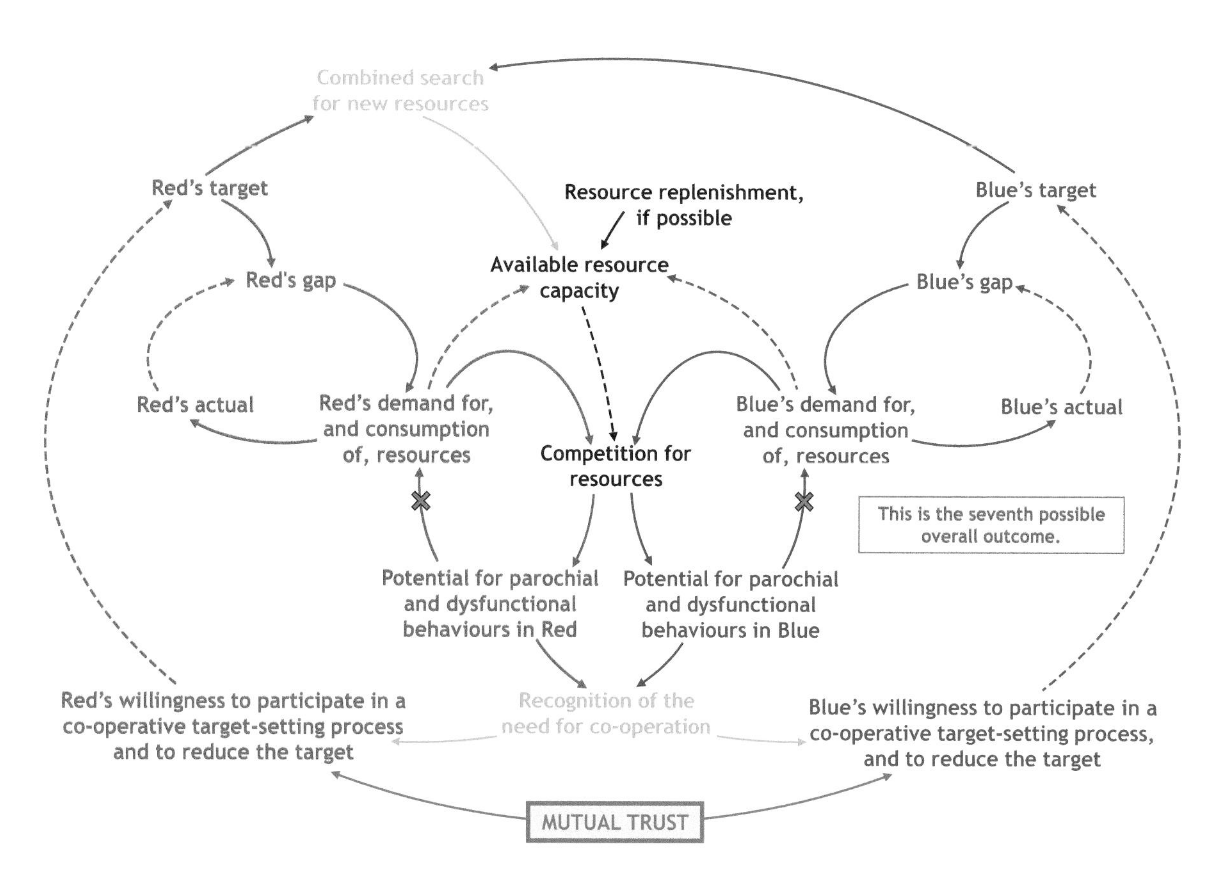

This is the seventh possible overall outcome.

But all this will happen only if there is a deep, and sincere, MUTUAL TRUST between Red and Blue.

Teamwork

Isn't that precisely what/teamwork is all about?

And is teamwork the emergent property of a community of people no longer 'doing their own thing' but behaving as a well-connected system?

The Story of 'Red' and 'Blue' Revisited

As the last several pages relate, the story of 'Red' and 'Blue' has seven, different, overall outcomes:

- *'happy ever after' (page 142)*
- *mutual destruction (page 148)*
- *'winner takes all' (page 148)*
- *conflict (page 153)*
- *appeal to higher authority (page 155)*
- *agreement to share the available resource (page 159)*
- *agreement to collaborate to discover new resources, if possible (page 163).*

Which outcome actually happens depends on the behaviours of, and choices and decisions made by, 'Red' and 'Blue'. Which outcome do you prefer?

- *Who do you think 'Red' and 'Blue' might be? How many different possibilities can you identify?*
- *Have you ever been 'Red' or 'Blue'?*
- *How did you behave? What were the outcomes?*
- *What might you do – possibly differently – when you are 'Red' or 'Blue' in the future?*

Chapter 12

Businesses Are Inherently 'Joined Up'

DOI: 10.4324/9781003304050-14

A Managerial Disconnect…

'In conclusion, the new marketing strategy will deliver significant profitable growth, and I ask for the Board's approval'. Alex smiled to everyone around the table. and sat down, confident the proposals would be accepted with acclaim.

'Just one question, if I may', said Sam. 'What delivery time are you promising?'

'48 hours, of course', replied Alex sharply. 'Like we always do'.

'In which case, I have a second question. What volume of sales are your expecting?'

'Well, obviously, an increase. About an additional 1,200 items a day'.

After a brief pause, Kim, the accountant, chipped in. '1,200? I've just done some calculations, and I can't reconcile that with the figure of the increase in profits you mentioned in your presentation'.

Alex, obviously caught off guard, looked at the presentation notes. 'Yes… er… I see… yes, an increase in volume of 1,200 is the low end of the range of estimates. The financial figures I presented were based on a higher increase to 3,000 a day'.

'3,000 a day!' exploded Sam. '3,000! No way can we deliver anything like that volume within 48 hours! We've been on the edge of our delivery capacity for several months now, and would be totally stretched at a volume increase of about 300, let alone 3,000! So you'd better tell the customers they'll be lucky for 96 hours!'

'96 hours! Don't be stupid! The customers will never accept that! They'll walk!'

Alex Is Focused on Marketing…

This is a representation of Alex's key interests and priorities.

Alex's *ambition is to grow* the business primarily by *marketing*, thereby progressively increasing the *customer base*, *revenue* and *profits*…

…so providing more *funds for investment*…

…into even more *marketing*…

…as shown by the resulting reinforcing loop.

Not all *profits* are available for *investment*, for there is a need to provide *returns to investors* too.

Accordingly, and assuming that the *returns to investors* have 'first call' on the *profits*, the *funds for investment* are the *profits* minus the *returns to investors*…

…and the *profits* themselves need to be in accordance with the agreed *financial targets*, shown in this diagram as a dangle associated with an influence link.

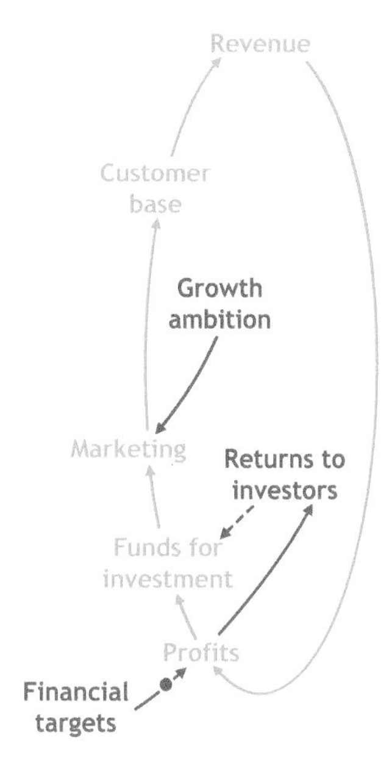

...Whilst Sam's Focus Is on Delivery...

Actual delivery time
Short-term action
Actual infrastructure

Sam's priorities are very different: success for Sam is to manage the *actual delivery time*, day in, day out.

The *actual delivery time* is determined by two factors, the first being the total capacity of the facility Sam is managing, encompassing machines, people, premises, transport..., all of which are collectively represented in this diagram as the *actual infrastructure*...

...and in general, the greater the *actual infrastructure*, the lower the *actual delivery time*, hence the inverse link.

The second factor is Sam's ability to take *short-term action*, as required: for example, to ask staff to work overtime, to hire temporary workers, to rent additional trucks... all of which act to increase the organisation's productive capacity, supplementing the *actual infrastructure*...

...once again acting to reduce the *actual delivery time*, or to prevent it from increasing, implying that the link from *short-term action* to *actual delivery time* is also inverse.

...Always Striving to Meet the Target Delivery Time...

Sam also has an important performance measure – the *target delivery time* – which the *actual delivery time* must never exceed. Since a short delivery time is 'good', this is represented by a balancing loop of the form shown on the upper half of page 73.

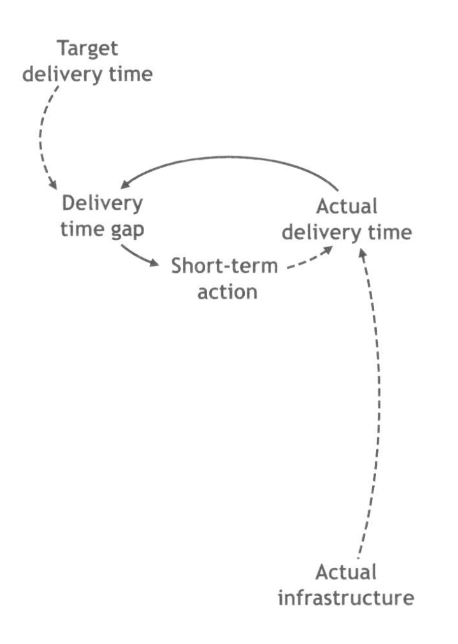

Sam continuously monitors the *delivery time gap*, the difference between the *target delivery time* and the *actual delivery time*, taking *short-term action* as required. And since 'small is good', it is more convenient to define the *gap* as

$$Gap = Actual - Target$$

hence the inverse link from *target* to *gap*, and the direct link from *actual* to *gap*.

If, for example, the *actual delivery time* is running at 40 hours, the *gap* is 40 − 36 = +4 hours. This (positive) *gap* drives a (positive) *short-term action*, such as working overtime in the shipping department, so that the *actual delivery time can be reduced* from 40 to 36 hours, as required.

That explains the inverse link from *short-term action* to *actual delivery time*, resulting in a closed feedback loop with one inverse link. One is an odd number, so this is, as expected, a balancing loop that acts to being the *actual delivery time* into line with the *target*.

If the *actual delivery time* were to fall to, say, 30 hours, the *gap* = *actual* − *target* = −4 hours, a negative number. This might be just ignored, with Sam feeling comfortable that he has some 'slack' available should things get busier; alternatively, it might trigger a 'negative' *short-term action*, such as working shorter hours, to allow the *actual delivery time* to rise a little.

…so Alex and Sam Live in Two Very Different Worlds…

As this diagram emphatically shows, Sam and Alex 'live in two very different worlds', which Sam and Alex, from their organisational – and parochial – viewpoints might be regarded as quite separate.

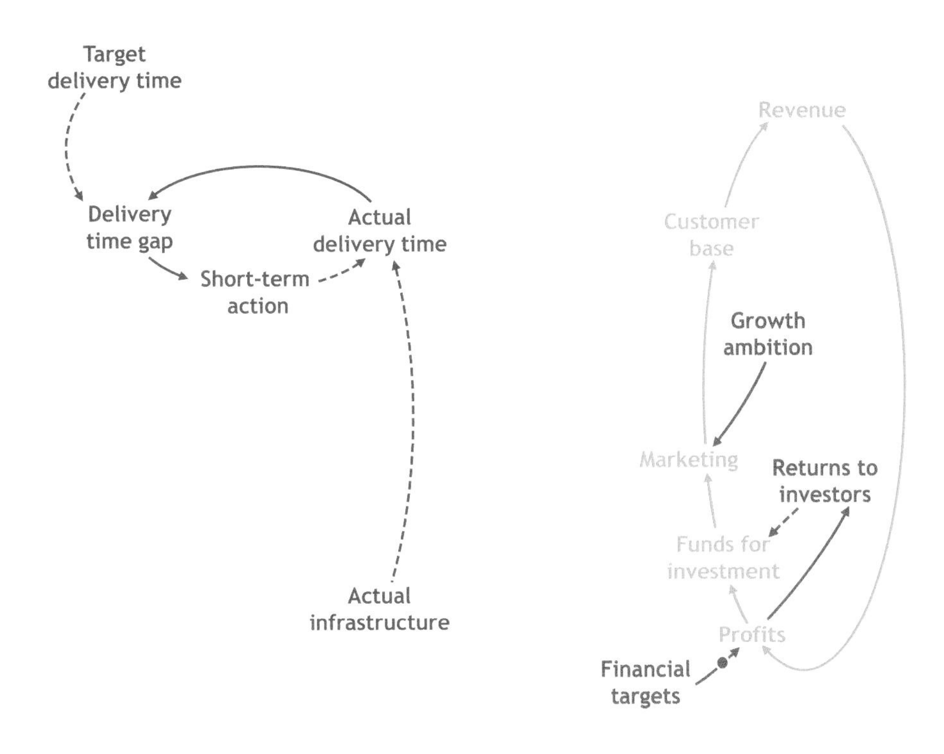

...but Which, in Reality, Are 'Joined Up'...

But the 'worlds' of Alex and Sam are in fact connected: if a *marketing* initiative is particularly successful (which is good), this will increase the *customer base* (which is also good), which, in turn, will generate an increase in the *actual demand*...

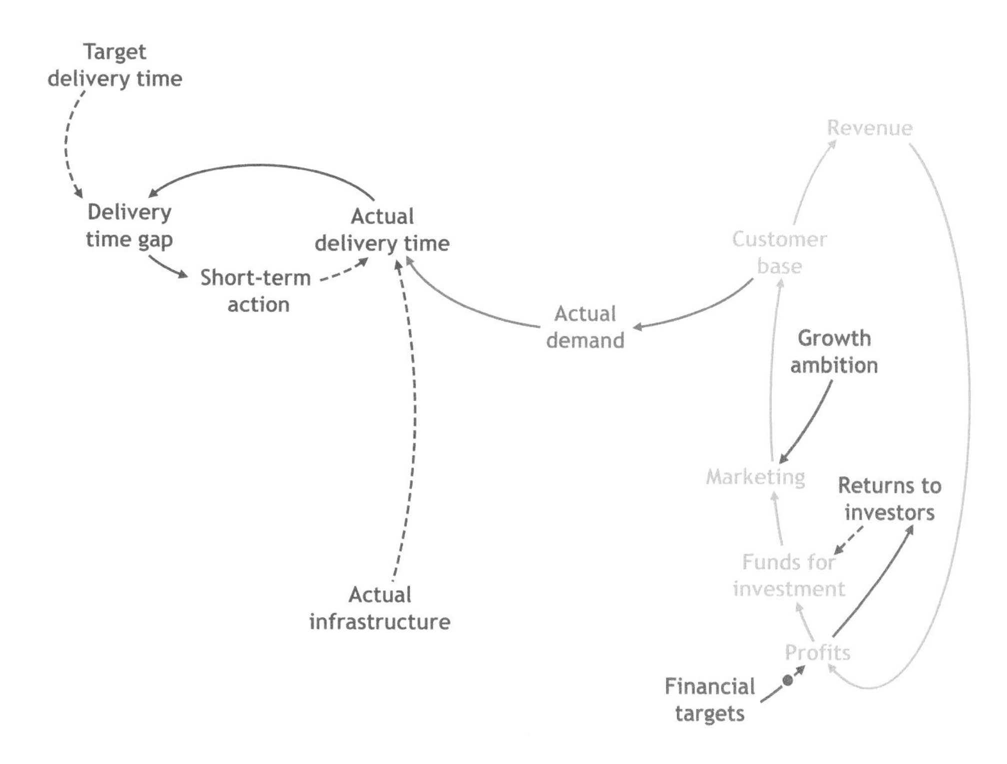

...which will, in turn, increase the production workload, and put upwards pressure on the *actual delivery time*...

...which Sam will manage by taking appropriate *short-term action*...

...but that has limits, for if the *actual demand* becomes even higher, the effects of the *short-term actions* that Sam can take can no longer keep the *actual delivery time* below the *target delivery time*.

...which is the situation described in the story on page 168.

An Important Balancing Loop...

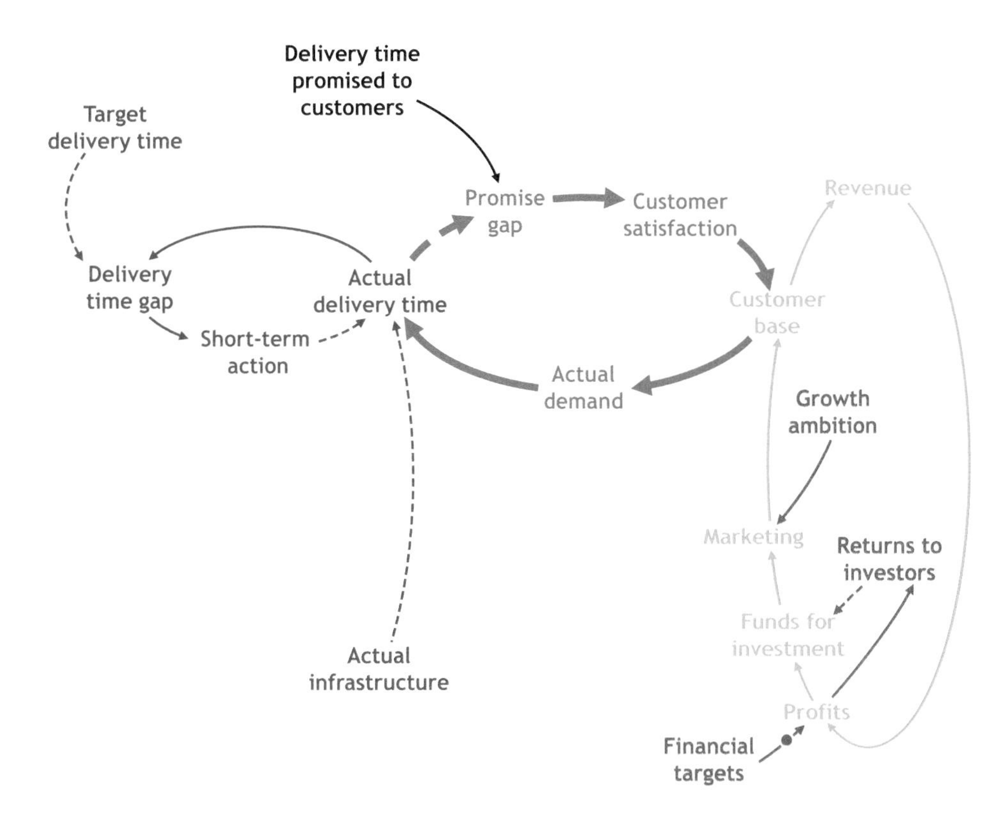

For narrative, see following page

...Which Can Turn Customers Away

Customers, however, have an expectation, perhaps in accordance with an explicitly stated, or inferred, *customer promise*...

...and any discrepancy between the *customer promise* and the *actual delivery time* will be noticed.

So if the *promise* is to deliver within 48 hours, and the *actual delivery time* is 28 hours, the *promise gap=promise – actual*=48–28=20 hours, a positive number. The *customer will be delighted* with the speedy service, which exceeded expectations, and perhaps will recommend a friend, so increasing the *customer base* even more.

But if the *actual delivery time* is 54 hours, the *promise gap=promise – actual*=48–54=–6 hours, a negative number, and the *customer might be very upset indeed*. That *customer* won't come back, and might even say 'don't go there' to friends. This diminishes the *customer base*, and, if severe, might even tip the customer reinforcing loop from exponential growth into exponential decline. Which is bad news indeed, possibly driving an exponential decline with a momentum that is very hard to arrest...

Even if that were not to happen, there is another effect, attributable to the closed feedback loop highlighted in magenta. This contains one inverse link, and so is a balancing loop which acts to limit the *customer base*, and hence the *demand*, to a level corresponding to an *actual delivery time* which is in line with the *customer promise*.

The business therefore ends up serving only those customers who are willing to wait.

Competitive Pressure

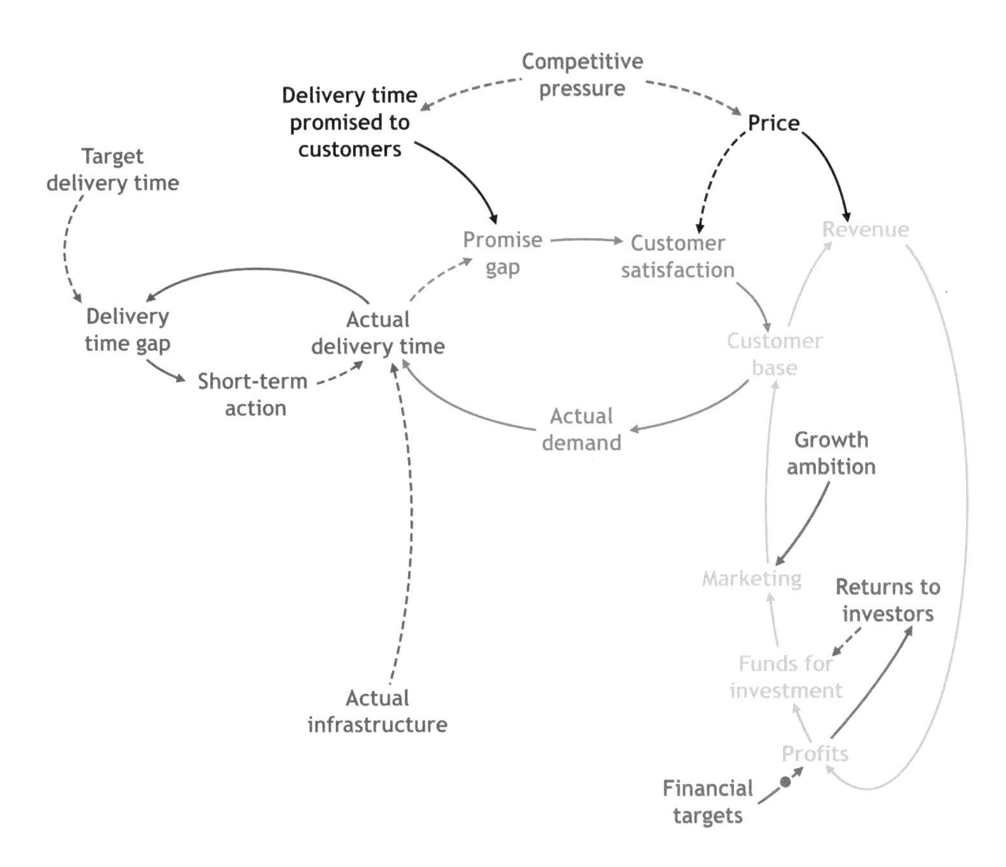

There are *competitors* out there too, and *competitive pressure* acts to force *prices* to be as low as can be tolerated, whilst still making *profits*; similarly, the *customer promise* must be for the shortest, quickest, possible *delivery time.*

Playing Safe

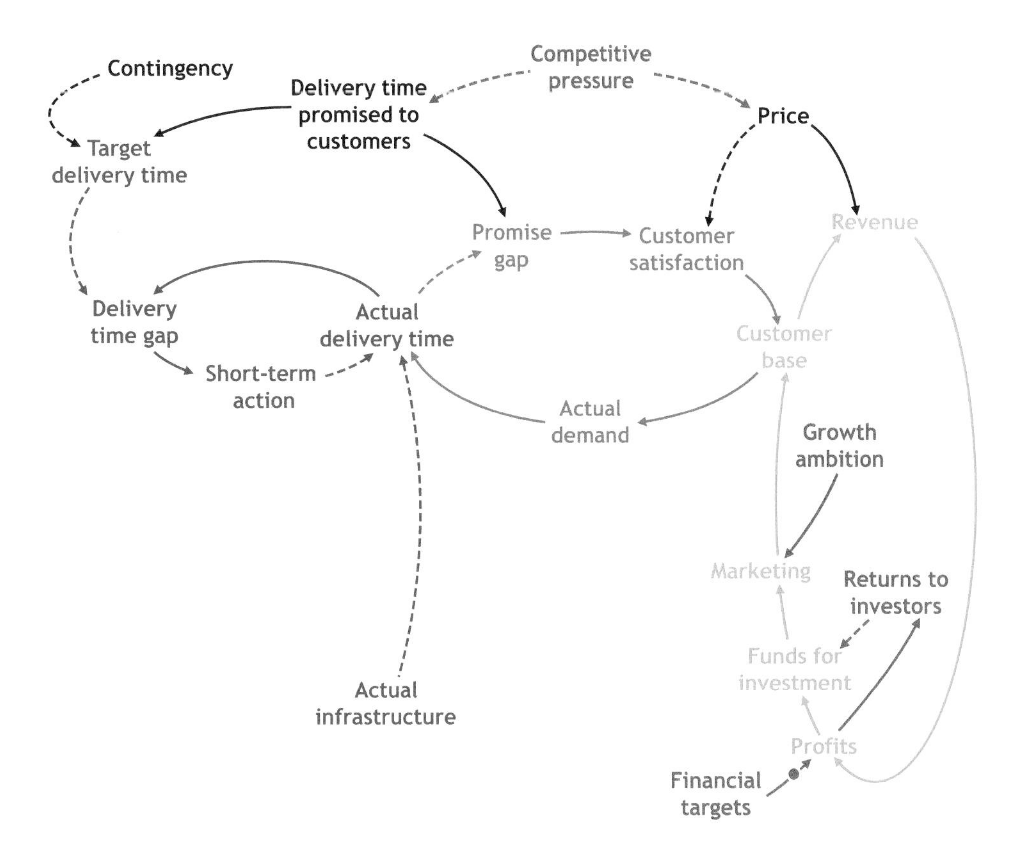

The *customer promise* also influences Sam's most important performance measure, the *target delivery time.* And a wise manager will build in some *contingency,* so that the internal *target delivery time* is even more demanding than the publicised *customer promise.*

So, for example, if the *customer promise* is delivery within 48 hours, the *contingency* might be 12 hours, giving a *target delivery time* of 36 hours. This builds in some 'slack', allowing for the internal *target* to be missed, for just a short amount, but without unduly damaging *customer satisfaction.*

The Infrastructure Balancing Loop

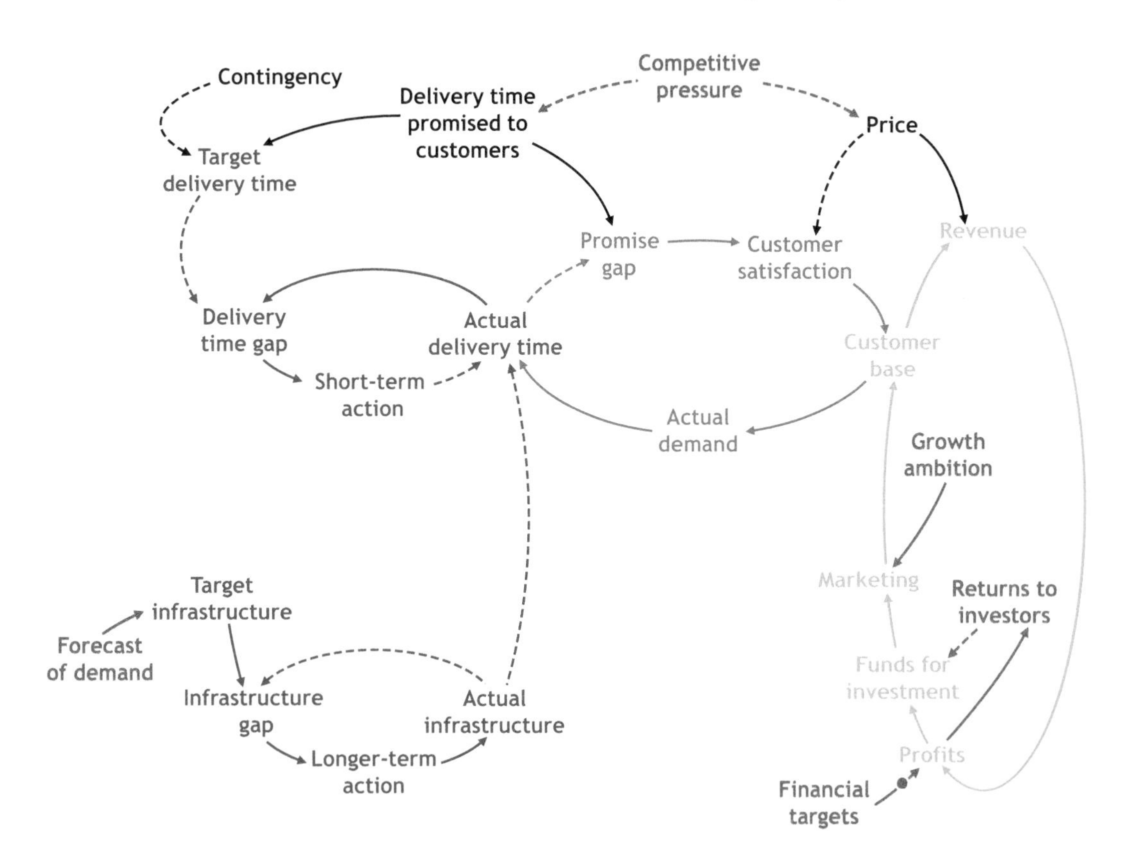

For narrative, see following page

The Infrastructure Balancing Loop

The *actual infrastructure* did not get built 'out-of-the-blue'; rather the *infrastructure as actually exists* is the consequence of a decision, taken at some time in the past, to build it at a size and scale that would be able to satisfy some future level of *demand*, as expressed, at that time, as a *forecast*.

In the language of systems thinking, an organisation determines a *target infrastructure*, envisaged to satisfy some *forecast of future demand*.

Initially, there is no *actual infrastructure*, and so that opens a significant *infrastructure gap = target infrastructure – actual infrastructure*.

This *infrastructure gap* triggers a building programme, which – especially by comparison to the *short-term actions* such as overtime – is inherently a *long-term action*, for the intent is to build an *actual infrastructure* that will last, ideally, for many years…

…so closing the *infrastructure gap*.

This forms another balancing loop, controlling the *infrastructure*.

Accordingly, if a decision were taken to expand the production capacity, this would increase the *target infrastructure*, opening the *infrastructure gap*, resulting in a *long-term action* of building, for example, a new warehouse, or commissioning new equipment, so increasing the *actual infrastructure*, with a possible consequence of reducing the *actual waiting time*.

Conversely, if a particular market is diminishing, a *forecast of future demand* might result in a decision to decommission some machines. This reduces the *target infrastructure*, leading to the (negative) *longer-term action* of removing the machines, so reducing the *actual infrastructure* to come into line with the lower *target infrastructure*.

Building the Infrastructure

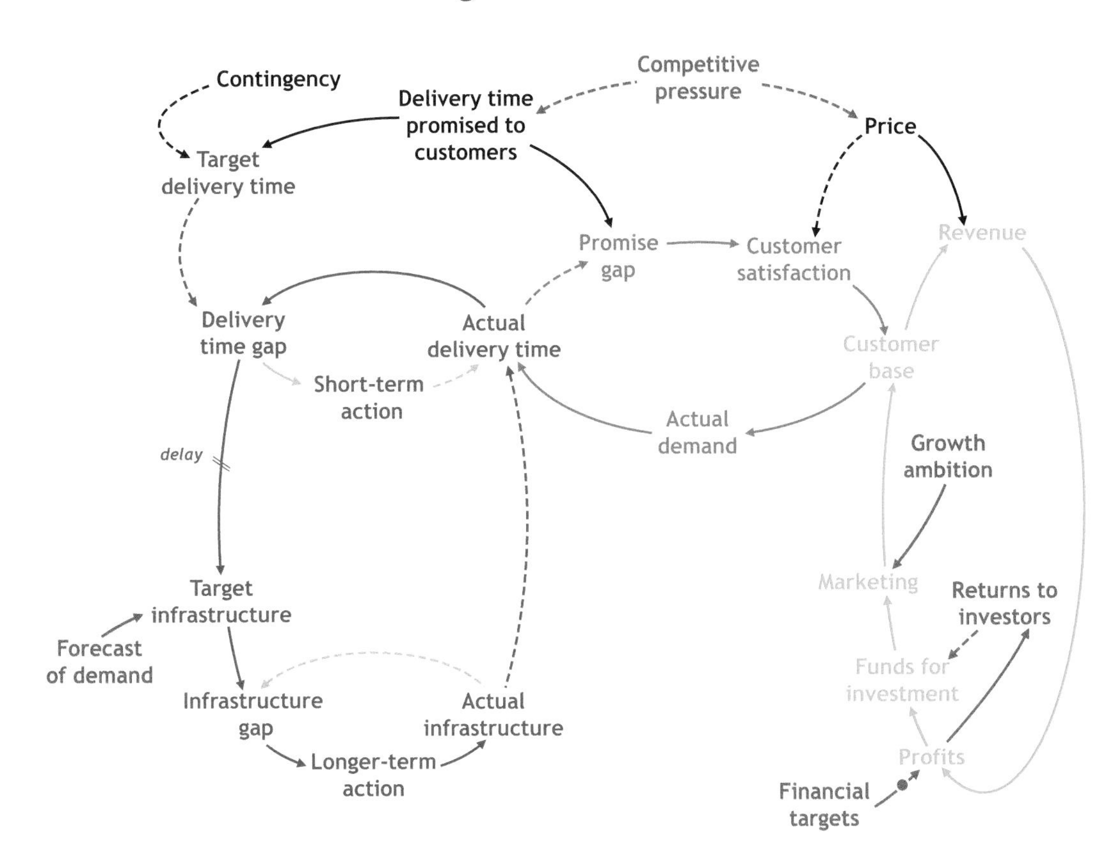

For narrative, see following page

Building the Infrastructure

The new feature shown on the causal loop diagram on page 180 captures what happens as Sam takes as much *short-term action* as possible, but is still unable to reduce the *actual delivery time* to the meet the *target delivery time.*

The fundamental problem is that the *actual infrastructure* just can't cope with the *actual demand,* even when supplemented by overtime, and temporary workers.

And so the fundamental solution is to increase the *actual infrastructure* on a more permanent basis. Organisationally, that implies the agreement of a larger *target infrastructure*, corresponding to a new *demand forecast* – a *forecast* validated by the increase in *actual demand* currently being experienced.

This introduces a new link from *delivery time gap* to *target infrastructure,* a link that is quite likely to be associated with a *delay* as Sam first tries as many *short-term actions* as possible, and then has to initiate what might be a lengthy organisational process to gain agreement to a project that might require substantial funds.

As can be seen in blue on the left, the direct link from *delivery time gap* to *target infrastructure* completes a balancing loop which acts to reduce the *actual delivery time* back down to the *target delivery time* – which is exactly what is required.

The Investment Choice

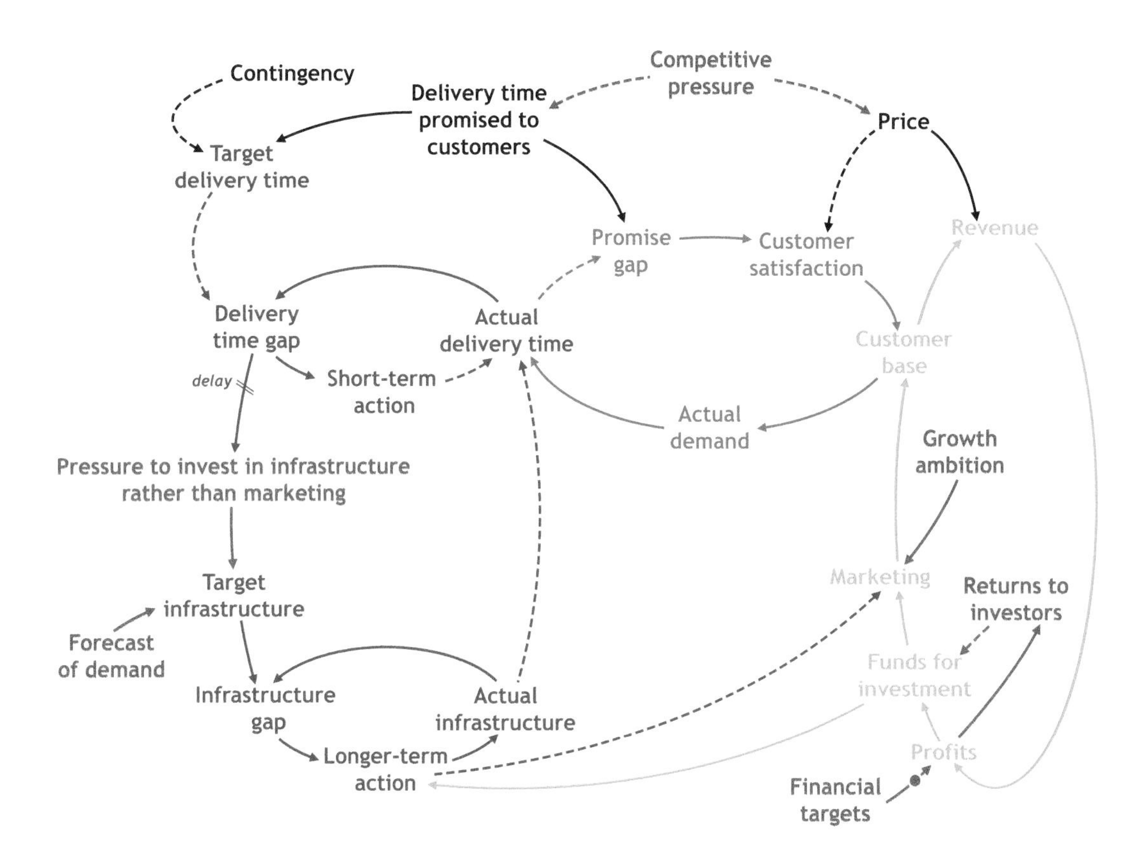

For narrative, see following page

The Investment Choice

The narrative on page 181 describes how Sam 'has to initiate what might be a lengthy organisational process to gain agreement to a project that might require substantial funds'.

The reality is likely to be more complex, and more political, than those words imply.

The funding that Sam requires to build a bigger target infrastructure is sourced from the *funds for investment*, which are also the source of the funds spent on *marketing*, as well as for any number of other possibilities not shown.

Accordingly, as the *delivery time gap* becomes progressively more adverse, Sam begins to exert internal organisational *pressure to switch investment funds from marketing to infrastructure*. The intent of this is to allocate appropriate *funds for investment* to the *longer-term action* – such as building or the purchase of equipment – required to enhance the *infrastructure*, as indicated by the direct link from *funds for investment* to *longer-term action*.

As a direct consequence, the funds allocated to *marketing* must simultaneously be reduced, as shown by the inverse link from *longer-term action* to *marketing*.

This inverse link indicates not only a reduction in the funding of *marketing*, but also that the investment in *infrastructure* is taking organisational priority, for it implies that

Investment in marleting = *Total funds for investment* – *Investment in infrastruture*

In essence, *marketing* is allocated whatever is left after the *investment in infrastructure* has been funded.

If the organisational priorities were the other way around, with *marketing* having 'first call', then

Investment in infrastruture = *Total funds for investment* – *Investment in marleting*

and the inverse link would be from *marketing* to *longer-term action*.

A Richer Picture

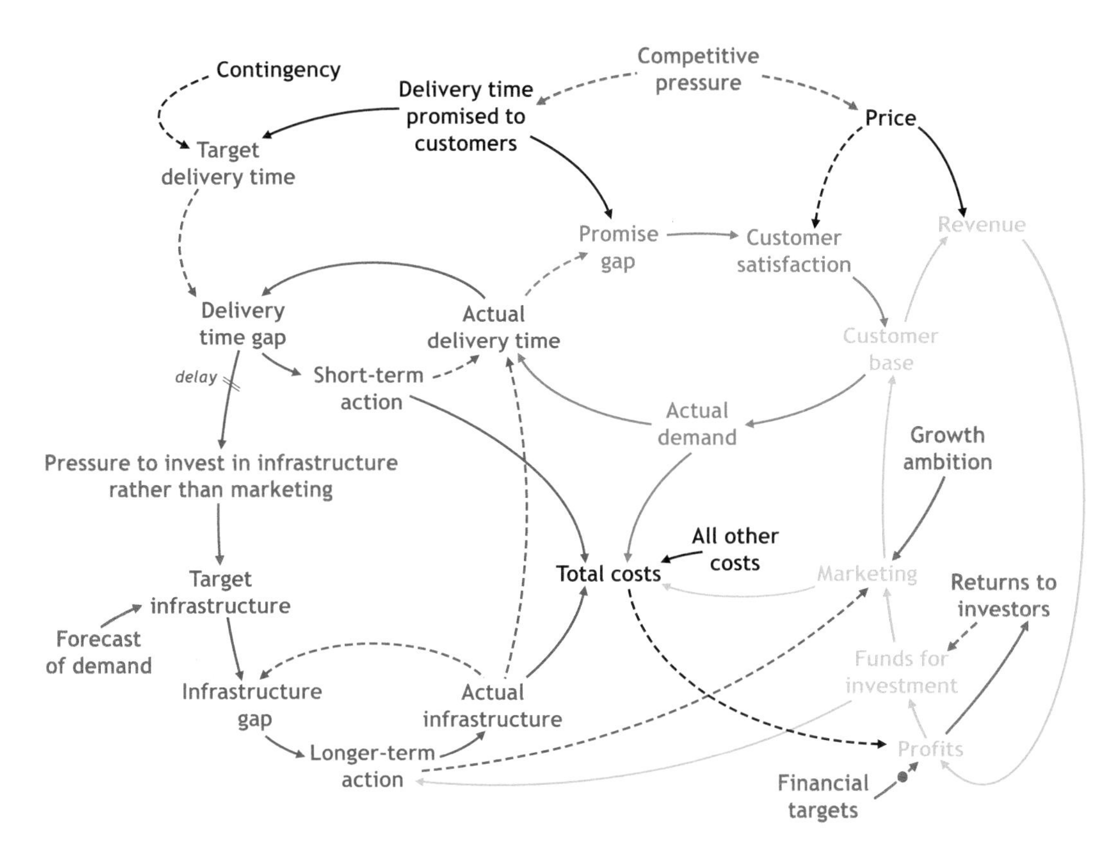

For narrative, see following page

A Richer Picture

The additional features in the causal loop diagram on page 184 identify some of the main contributors to the organisation's *total costs.*

Although financial items can be included in a causal loop diagram, the purpose is not to duplicate a financial spreadsheet. So the number of financial items typically shown is fewer rather than more, and the items that are shown tend to be more general rather than detailed.

So, for example, the *total costs* are just that, and deducted from *revenue* to determine the organisation's *profits,* with the major contributors to the *total costs* being the operating costs associated with

- marketing
- meeting the actual demand, these being, for example the total costs of production...
- short-term action, such as overtime payments...
- the actual infrastructure, such depreciation, maintenance, insurance...
- plus the catch-all of all other costs.

The Reinforcing Loop of Growth

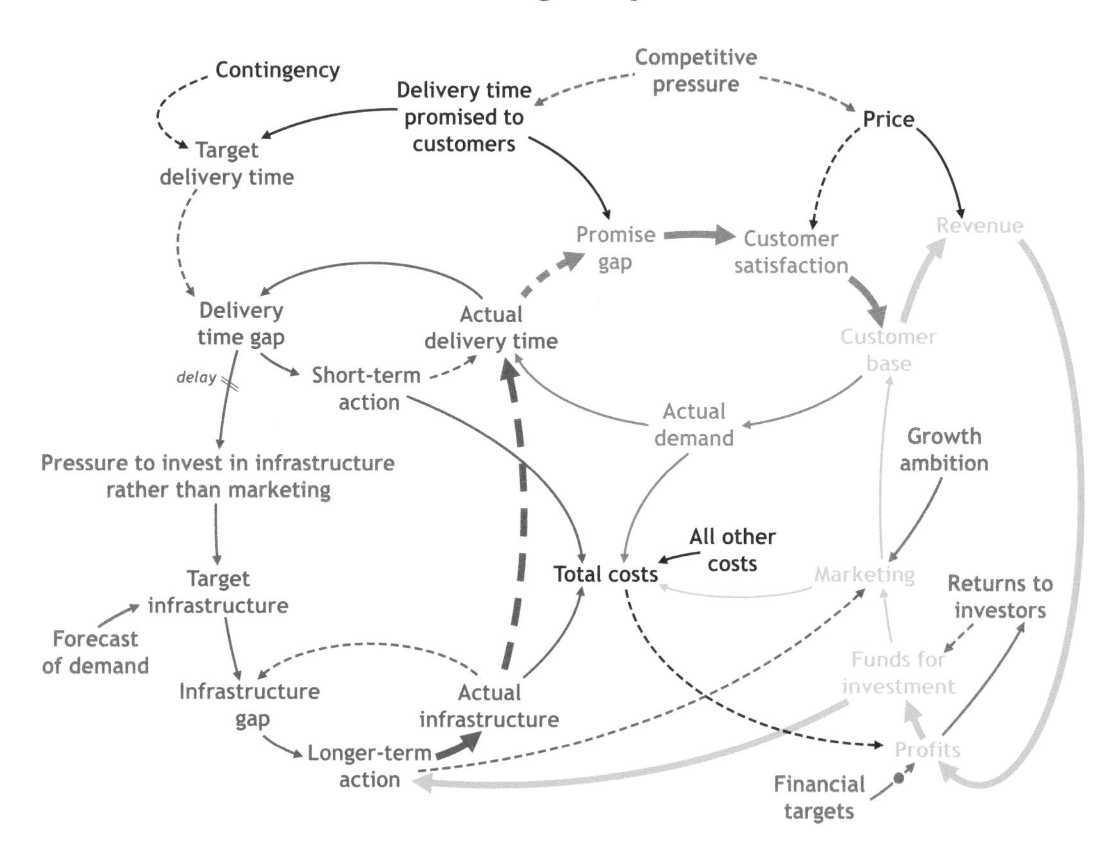

For narrative, see following page

The Reinforcing Loop of Growth

A (not very obvious!) feature of the causal loop diagram shown on page 184 is the closed feedback loop, as emphasised on page 186, from *revenue* and *profits* to *funds for investment*, then to *longer-term action, actual infrastructure* and *actual delivery time*, and finally to *promise gap, customer satisfaction* and *customer base*, back to *revenue*.

Though somewhat strangely shaped, this sequence is indeed a closed feedback loop, comprising nine links of which two are inverse – these being the links from *actual infrastructure* to *actual delivery time* and from *actual delivery time* to *promise gap*.

Two is an even number, so this is a reinforcing loop which will show exponential growth or exponential decline.

Of those two, exponential growth is the more appealing – but what does this loop mean?

In a business context, this is about infrastructure-led growth in which a steady *investment* in *infrastructure* drives a progressive reduction in the *actual delivery time*, consistently exceeding customer expectations, delivering continuous and sustained *customer satisfaction*, so growing the *customer base*.

A System to Avoid

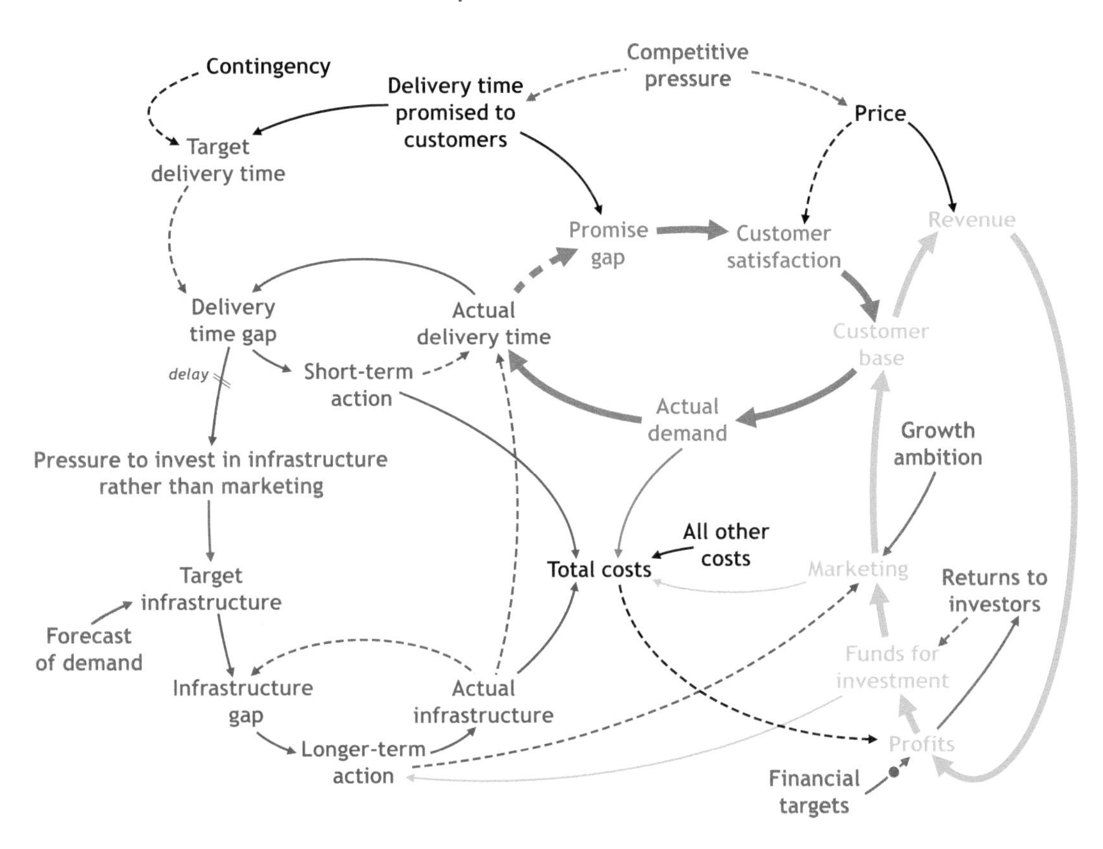

For narrative, see following page

A System to Avoid

The causal loop diagram on page 188 highlights the *marketing-driven* reinforcing loop, and also the associated *demand – customer satisfaction* balancing loop.

If the *actual demand* grows too quickly, this can overwhelm the *actual infrastructure*, driving an increase in the *actual delivery time*. This opens the *promise gap*, driving down *customer satisfaction* and reducing *demand*, which, commercially, is not a good place to be...

The diagrams on pages 186 and 188, however, are identical, as is the underlying system.

How the system actually behaves in practice, and whether the outcome is sustained growth or a *BIG PROBLEM*, depends wholly on how the organisation's management take those all-important decisions concerning the allocation of the enterprise's *funds for investment*.

The Best of Both Worlds

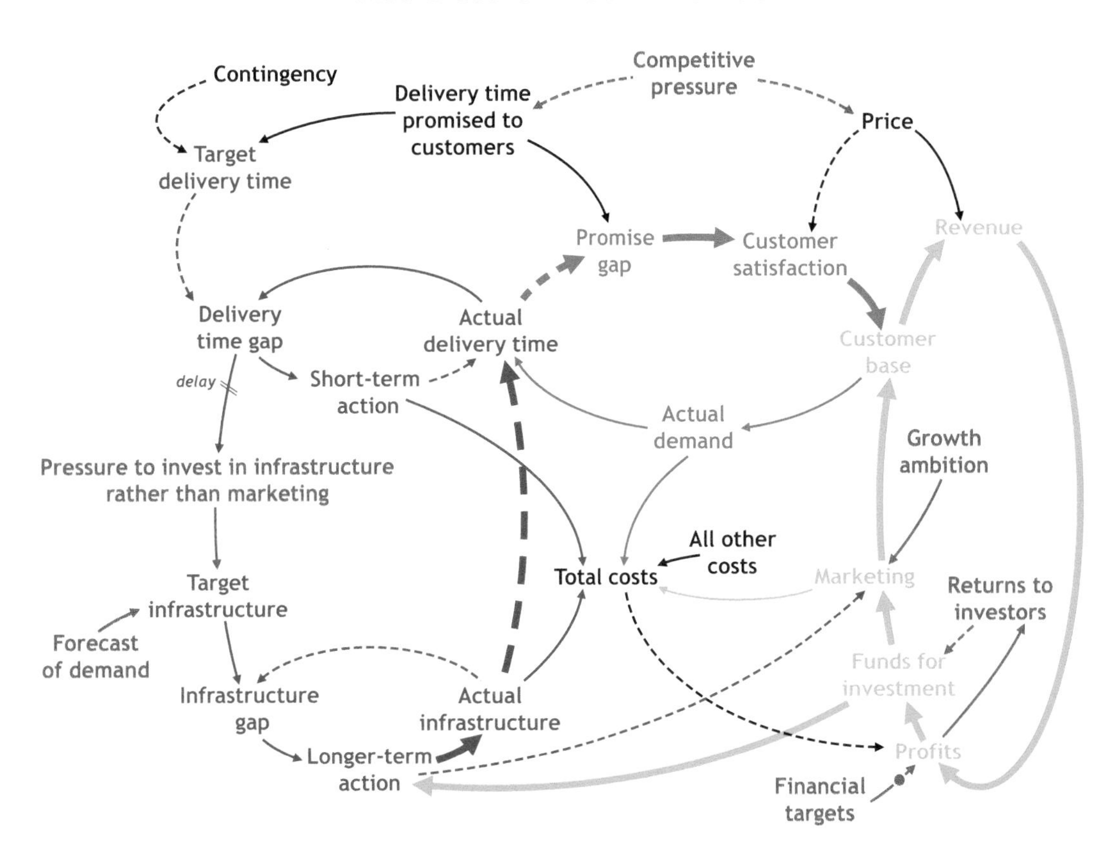

For narrative, see following page

The Best of Both Worlds

The strategic 'sweet spot' is to run both reinforcing loops, in harmony, together, so that the *infrastructure* is always just ahead of a steadily increasing *customer demand* resulting from a progressive building of the *customer base*, as achieved by a well-managed blend of effective *marketing* and *infrastructure-driven customer satisfaction*.

Easier said than done, for actually accomplishing this – as every manager knows well – is exceedingly difficult. Not only does it require the 'joining up' of all the relevant management processes, but also the 'joining up' of the management team too, so that there are no 'disconnects' such as that at the heart of the story related on page 168.

And having a causal loop diagram, such as that shown on page 190, can surely help, for the business is intrinsically joined-up.

This causal loop diagram is complex – which is no surprise, for the business is complex. But this complexity can be tamed, for the causal loop diagram, though indeed complex, is intelligible. Especially since, as we have seen, the description of the system is just a network of interconnected reinforcing and balancing loops.

The diagram also highlights the key strategic decision – the allocation of *investment funds* between *marketing* and *infrastructure*.

This binary choice is a simplification, for as discussed on pages 94 and 95, all real enterprises have more than two opportunities for *investment*. But that allocation decision remains *THE* key management decision, and an understanding of the appropriate causal loop diagrams – such as those shown on pages 94 and 190 – can be enormously helpful in supporting the managerial discussion, and subsequent decision.

Even better is to quantify what is happening, which is about using the causal loop diagram as the specification for a system dynamics computer simulation model, as discussed on page 312.

A Frequently Met 'Building Block'

Although the causal loop diagram on page 190 has been described in terms of a business that despatches products to *customers* within a *promised delivery time*, the diagram has very wide applicability. This may require changes to some of the words. and – for many public sector organisations that do not explicitly sell anything – changes to the *revenue* reinforcing loop to recognise other ways in which *demand* both arises of its own accord (for example, by the natural growth of populations), or can be stimulated (for example, by advertising to encourage the use of particular services).

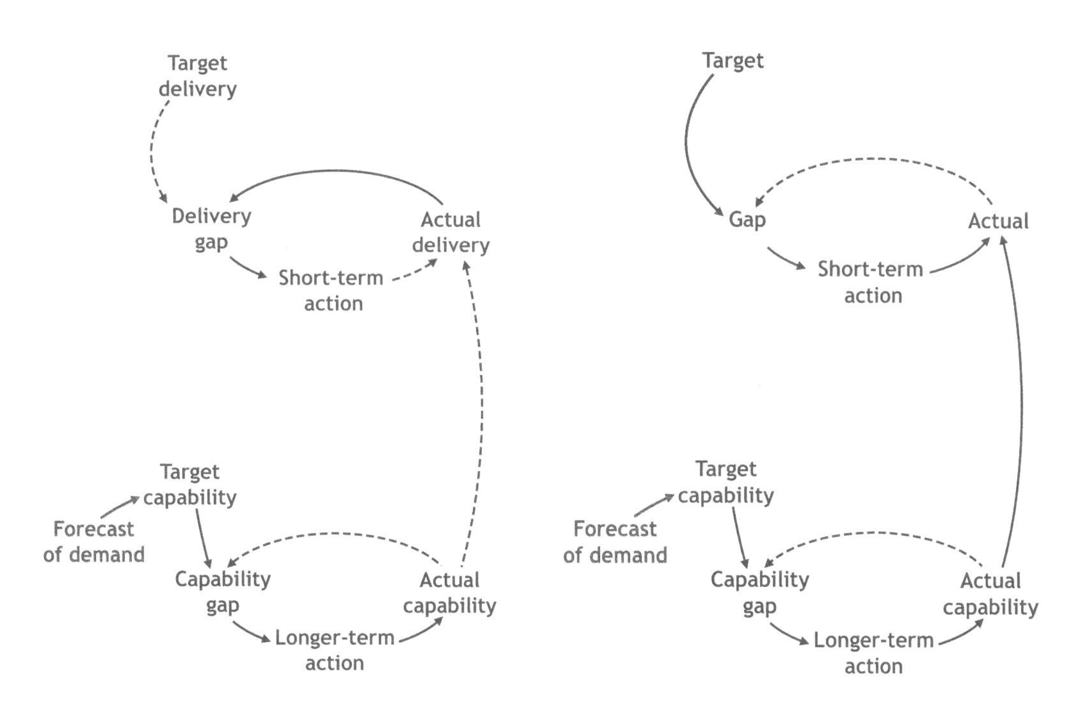

Furthermore, the structure represented by the causal loop diagrams to the left are ubiquitous, and apply wherever 'something' is delivered, with that 'something' being created by an appropriate 'capability'. For example, the 'delivery' might be medical diagnostics, with the 'capability', the number of MRI scanners; the 'delivery' might be consulting services, and the 'capability' the number of consulting staff; the 'delivery' might the speed of a supermarket checkout, and the 'capability', the number of tills that are open.

One point to bear in mind: the form of the upper balancing loop, as shown to the far left, is appropriate for *targets* for which 'small is good' such as delivery times. For *targets* for which 'big is good', such as 'patients treated per week', the alternative form of the balancing loop, as shown in the centre, is often easier to use.

Chapter 13

Targets and Budgets in Practice

DOI: 10.4324/9781003304050-15

'My Job Is to Treat Patients...'

'You've overspent your medications budget'.

'And?'

'Well... er... you're not supposed to do that'.

'My job is to treat patients. Not to be some sort of petty financial bureaucrat!'

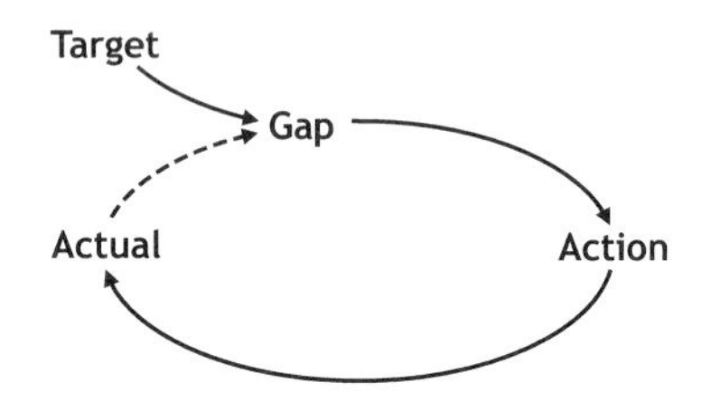

Well, that's clear! It seems as if the doctor hasn't attended a business school, and has therefore not been brainwashed into believing that overspending the medications budget is the gravest of sins from which there is no redemption.

Oh dear. Nor is the doctor complying with the balancing loop description (see page 70) of how budgets and *targets* are supposed to work.

In this context, the *target* is the original budget, this being organisational approval to spend [this much] on medications over the financial year; the *action* is the prescribing of medications; the *actual* is the corresponding cumulative cost; and the *gap* is the amount of money still available within the original budget for the rest of the year.

Managerially, monitoring the *gap* is intended to signal to the doctor to 'be careful', and adjust the level of prescribing so that by the end of the year, the *actual* spending on medications is equal to the *original budget*, so that the *gap* at the year end will be zero. But that hasn't happened. The *actual* spending on medications has exceeded the *original budget* with the result that the *gap* has become negative – that's the overspend. But the doctor doesn't care.

This reality is not recognised in the causal loop diagram, so this chapter takes a deeper look at how real balancing loops actually operate in practice...

The 'Theory'...

The starting point is this balancing loop, which corresponds to any context in which the *target* is lower than the *actual,* so that that the *gap* is always a positive number. This drives whatever *action is appropriate* to decrease the *actual* towards the *target* until such time as the *gap* becomes zero, at which point the loop switches off. This therefore relates to any situation in which 'small is good', for example a system associated with a target expressed as a (short) time to deliver a service, as discussed on page 171. I'll deal with the alternative case – in which 'big is good' – on page 203.

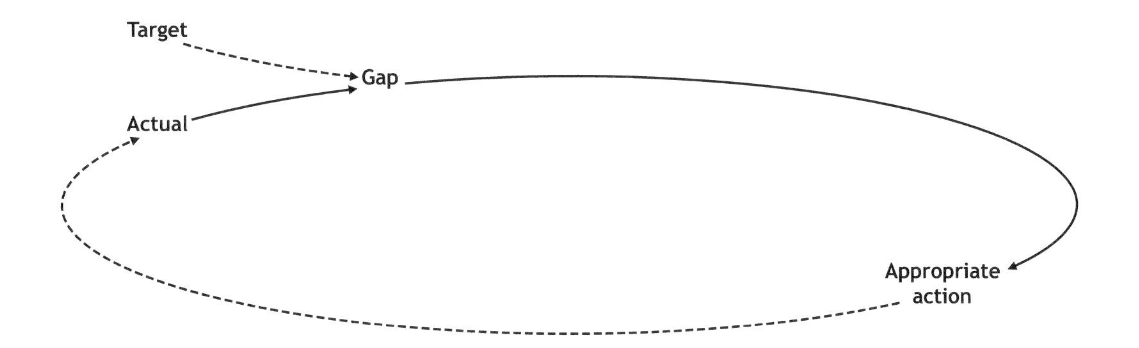

This causal loop diagram implies that any non-zero *gap* will trigger a corresponding *appropriate action*, and do so immediately. This is what happens in engineering control systems, such as thermostats, and also in managerial systems in which the *target* is not contentious, and where the *appropriate action* is well-understood and (relatively) easy to carry out.

But as the story of the doctor described, if the *target* is disputed, things can be very different...

...but 'Real Life' Probably Looks More Like This...

...for the reality is that the psychology of *targets*, *actuals* and *gaps* assumes that the existence of a *gap*, of whatever magnitude, will trigger the *motivation*, or perhaps a state of *anxiety*, resulting in a *pressure to act*, which, in turn, leads to the *appropriate action*.

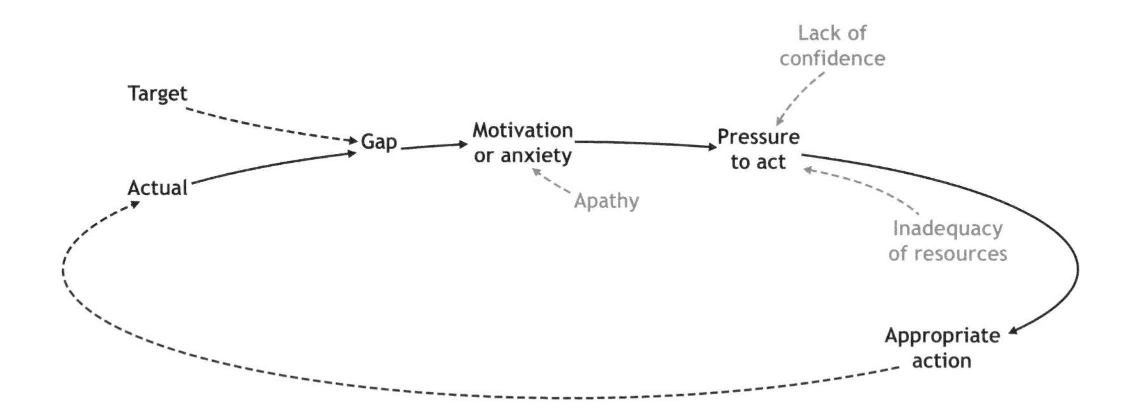

According to 'business school theory', *motivation* or *anxiety* are immediate reactions to any *gap*, and the *pressure to act* is intrinsic to any conscientious manager, so these two 'intermediate states' are often ignored. But they can often be real. For example, if the individual who is supposed to respond to the *gap* just *doesn't care*, then the *gap* can be as large as you like, but nothing will happen. Think of the teenager told to 'tidy your room' by a parent!

And even if the individual who is supposed to respond to the *gap* does *care*, that person might still be very *reluctant to act* if he or she *lacks confidence* in being able to take whatever the *appropriate action* might be, or *does not have the resources required* for successful delivery.

Patients Left for Hours in Ambulances

If the *appropriate action* is not taken in response to a *target*, things can go badly wrong. As an example, this is what happened at some hospital Accident and Emergency ('A&E') departments in England following the government's introduction, in 2005, of a performance measure that at least 98% of patients arriving at an A&E department must either be satisfactorily treated and discharged, or transferred elsewhere within the hospital, within 4 hours.

At that time, the average actual waiting time was considerably in excess of 4 hours, with wide variations around the country.

The intention was therefore for this target to act as an incentive for hospitals to improve the efficiency of their services.

The reality, however, was rather different...

> ***Ambulances, crews and patients are being deliberately delayed, often for hours, at hospital accident and emergency departments in order to meet government treatment targets. With a commitment to deal with all casualty cases within four hours of their arrival at a hospital, executives have been forced to keep patients waiting in ambulances until their staff can deal with them in the allotted time.***
>
> ***The Observer, 17 February 2008***
>
> *https://www.theguardian.com/society/2008/feb/17/health.nhs1.*

That's taken from an article in the newspaper *The Observer,* published on 17 February 2008.

The 4-hour target is measured from the time a patient 'arrives', this being recorded in a log book at the reception area. So any patients in ambulances parked outside have not yet 'arrived', allowing the 4-hour target to be met...

This practice became widespread, and was known as 'ambulance stacking'.

'Gaming' the System

So far, I've assumed that there has been only one way of taking the *appropriate action*. Sometimes, there is only one way, but in many contexts, there are many possible ways in which a *target* can be achieved, some '*harder*' than others.

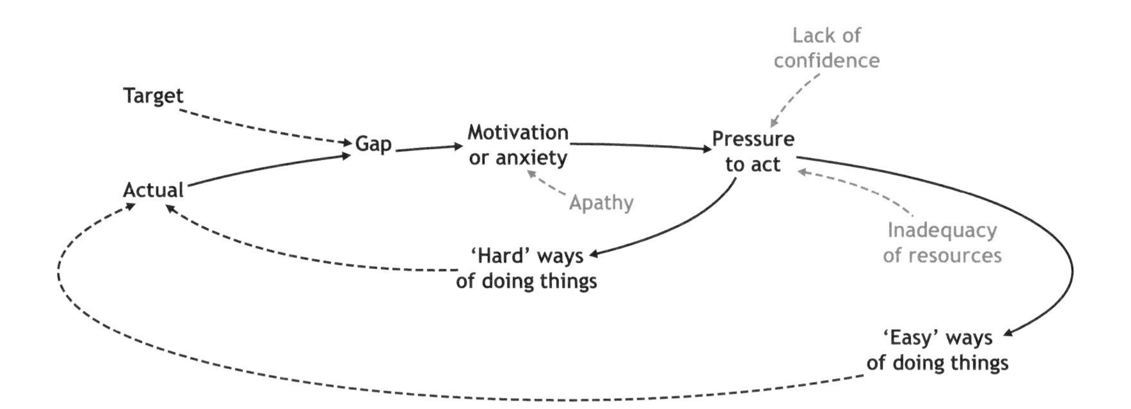

So, for example, in the (true) story of the hospital A&E department, for which a 'small waiting time it good', the '*hard ways*' of meeting the *4-hour target* are about improving the efficiency of the department, which probably requires improvements to the efficiency of many other aspects of the hospital's activities too. That is indeed *hard*, and is likely to take a long time to accomplish.

And whilst all this *hard work* is being carried out, the *4-hour target* will not be met, and the hospital management will be regarded as having failed.

By contrast, one example of an '*easy way*' is to leave patients outside in ambulances, and delay their 'arrival' until it's quite certain that the *4-hour target* will be met…

So What Do I Do?

In general, the *'hard' ways* are the 'right' ways, and what the 'target setter' – the person who sets the *target* and who wants something to happen – intended, whilst the *'easy' ways* are often 'cheats' or 'games' that wily 'action takers' – the people who actually take the *action* – discover, and often get away with. Given that the action taker has *a choice of which path to follow*, if *motivated* to take any path at all, the target setter might introduce *incentives* to counter any *apathy* and so increase *motivation*, and to encourage action takers not just to *take action*, but to take the 'right' and usually '*hard*', *ways* of doing things, whilst *disincentivising*, if not penalising, the *'easy' ways* of cheating and game playing, as will be discussed further on page 206.

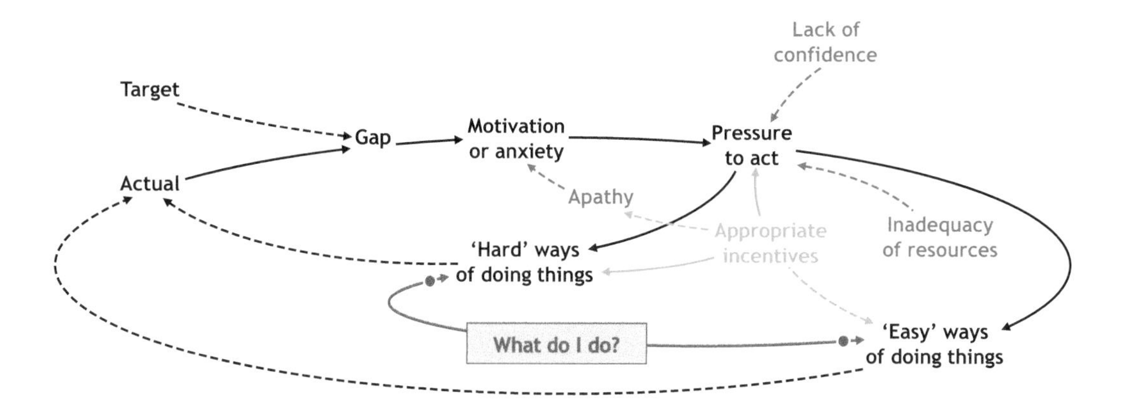

When the *'easy' ways* are discovered, the target setter often refers to **unintended consequences**, which is usually code for 'don't blame me about this – no one could possibly have imagined that would happen'.

No.

It's always possible to anticipate such events. Whenever a performance measure is set, the possibility that some people will attempt to cheat is immediately on the table. So asking 'how many ways can we think of to cheat this performance measure?', right at the outset, will identify some, if not many, of the possibilities, enabling them to be designed out, as will be explored on page 230.

But Suppose I Think the Gap Is Too Big?

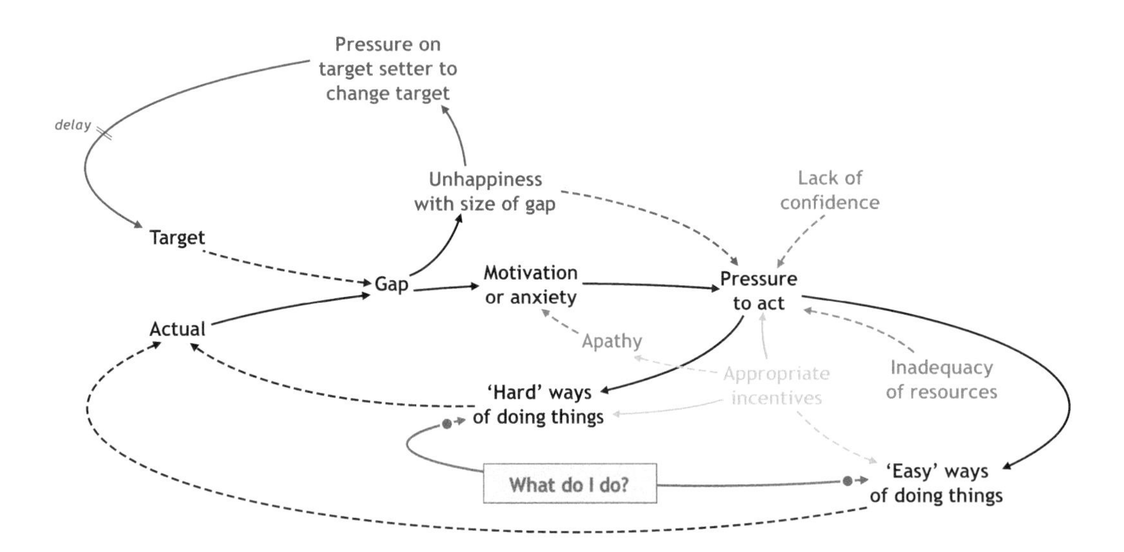

It might happen that someone given a *target* to achieve feels that it is unreasonable, in that the *gap* is just too big. A factory manager, for example, will surely agree that customers should not have to wait a long time for their orders, but might feel very concerned about the imposition of a next day delivery *target* when the current *actual* is four days. Any fear that the gap cannot be closed might reduce the *pressure to act*, for the task demanded is 'mission impossible'.

In many contexts, the power difference between the target setter and the action taker is such that the action taker might be *unhappy*, but can do nothing. Sometimes, however, it might be possible for the action taker to exert *pressure on the target setter to change the target* to make it easier to achieve, which might result in a 'softening' of the *target* (for example, in the case of hospital A&E departments, to increase the *target* from four hours, to, say, five, hence the positive link), perhaps after some *delay*, as determined, for example, by the amount of time that the 'negotiations' might require.

Or Disagree with the Target Itself…

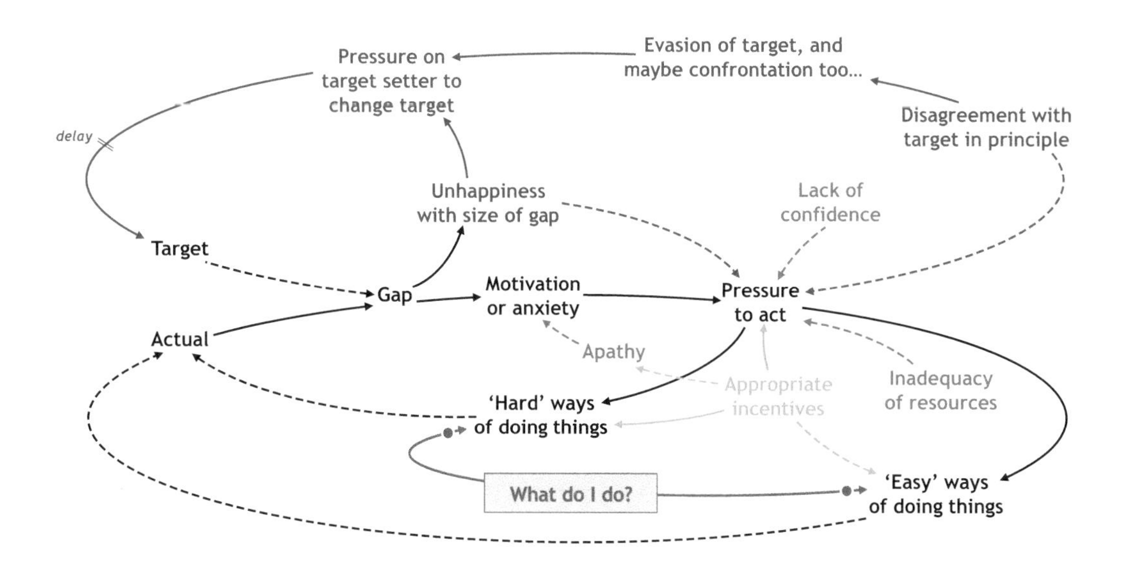

Another possibility is a *disagreement with the target in principle*: the action taker just believes that it is the wrong thing to do. This too reduces the *pressure to act*, possibly to the extent that the *target is ignored or evaded altogether*, so no action is taken at all, not even the *'easy'* ones.

The action taker's inaction will, sooner or later, be noticed by the target setter, which might result in some form of *negotiation* or even *confrontation* in which the target setter seeks to persuade the action taker to do whatever is required, whilst the action taker seeks to *persuade the target setter to 'soften' the target*, or change it more radically.

...Which Might Result in Conflict...

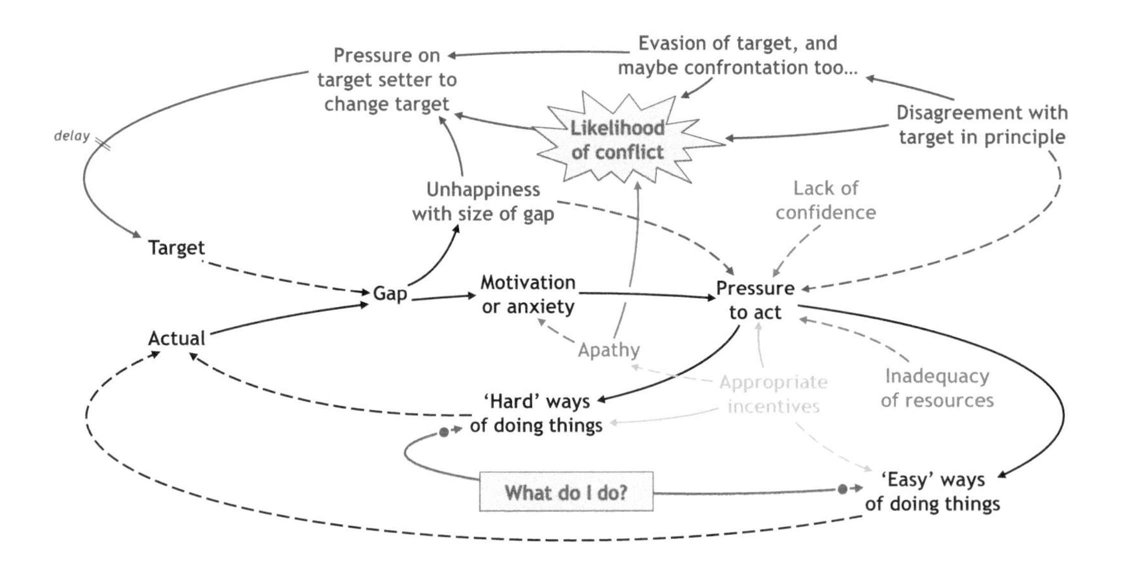

Any *confrontation*, however, might escalate into *conflict* – as indeed might happen as a result of the action taker's *apathy*, or a more active *disagreement with the target in principle.*

This explains, for example, many industrial strikes. In this case, the *actual* might be the current pay and conditions, the *target* might be management's proposals, and the *gap* might be, for example, a change to overtime payments or perhaps the pension scheme. In the view of the management, *anxiety* would be, say, the fear that workers might lose their jobs, driving a powerful *pressure* to take *action to sign up the new pay deal*, so bringing the *actual* into line with the proposed *target*. From the point of view of the workers' trade union, *disagreement* will drive *conflict,* with the threat of a strike – or an actual strike – putting *pressure on management to make the proposed new pay deal more favourable to the workers.*

The Diagram When 'Big Is Good'

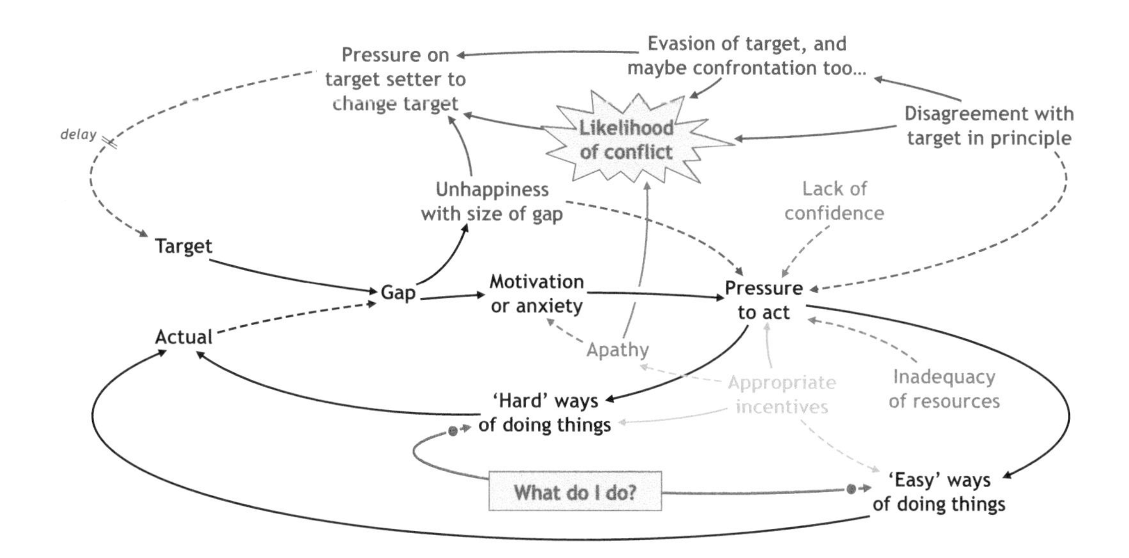

This is the diagram for a context in which 'big is good', for example, when the *target* is a performance measure, as applied to schools, such as 'the percentage of students awarded top grades in public exams', as will be described on pages 208–217. When 'big is good', in general, the *actual* will be smaller than the *target*, resulting in a positive *gap,* which, in turn, drives *'easy' ways* and *'hard' ways*, both of which have the effect of increasing the *actual*, hence the direct links.

Also, any *pressure on the target setter to change the target* has the intent of making the *target* lower, and therefore easier to achieve, hence the inverse link.

But If the Gap Just Doesn't Get Closed…

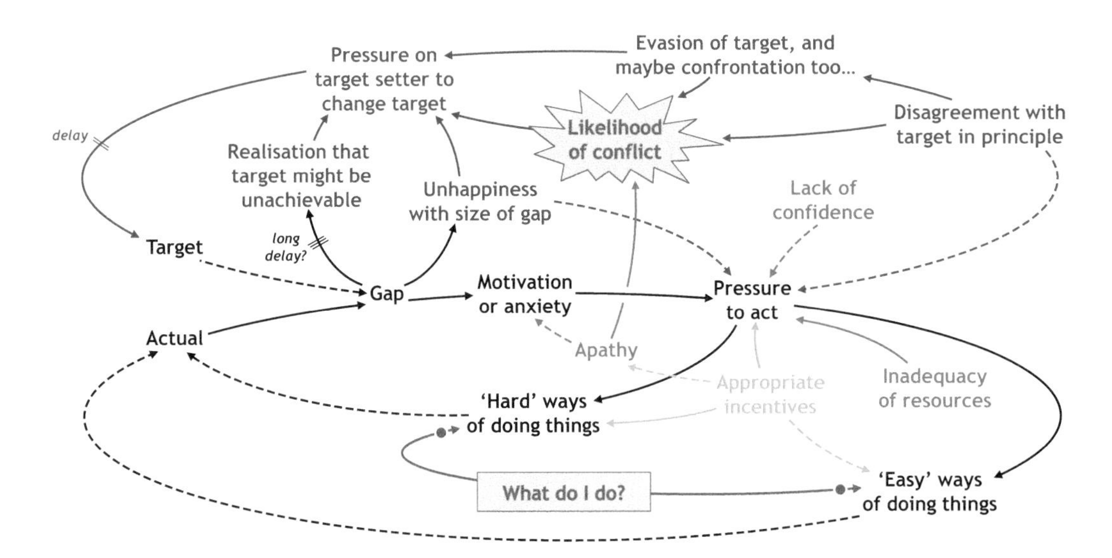

Sometimes a *gap* remains, no matter how hard conscientious people work to close it. One 'senior management' response might be to fire the 'middle manager' who, apparently, wasn't trying hard enough, only to find that the replacement, and the replacement's replacement… can't close it either. At which point the *penny might drop that the target is unachievable*… The prolonged existence of the *gap* therefore itself exerts *pressure on the target setter to change the target*, usually after a (possibly very) *long delay*. The example shown here is for 'small is good', as, for example, the 4-hour waiting time for admissions to hospital Accident and Emergency departments, as featured in the story on page 197 – which has a sequel, as briefly described on the next page…

...Sometimes 'Target Setters' Can Change Their Minds

NHS ENGLAND TO SCRAP 4-HOUR A&E TARGET AND REPLACE WITH AVERAGE WAITING TIME

New bundle of measures overwhelmingly supported by health professions in what will be the first change to system in 15 years.

The i newspaper, 26 May 2021

https://inews.co.uk/nhs/nhs-england-scrap-four-hour-ae-target-replace-with-average-waiting-time-1020912.

In 2000, the UK government set a target that, by 2004, no patient should wait longer than 4 hours after arrival at the Accident and Emergency Department in any hospital in England.

This was not achieved, so in 2005, the target was softened to 98% of patients to be treated within 4 hours rather than all patients.

This too was not achieved, so the target was reduced to 95% in 2010.

As this newspaper clipping shows, in May 2021, NHS England – the senior government body running the heath service in England – announced its intention to replace the 4-hour target by a 'bundle' of 10 metrics, including targets for the percentage of patients transferred from ambulances to the A&E Department within 15 minutes (so addressing 'ambulance stacking'), and measures of the average time for different categories of patient to remain within the A&E Department.

That's an example in which the target setter is someone different from the person, the action taker, who has to deliver the target – and it is by no means unusual that the target setter can be very resistant to any pressure from the action taker for that target to be changed, no matter how long the *gap* might last: in this example, it took 15 years. But when you are setting your own target – such as 'this month I must lose some weight/start learning another language/...' – the pressure to change the target can act much more quickly ('ah well, maybe next month...')!

Target Setters, Action Takers, Incentives and Nudges

A target setter who has some form of managerial authority over the action taker has the power to determine incentives that directly encourage the 'right' actions (such as a cash bonus to an employee), and also disincentives that deter the wrong ones (say, the threat of dismissal), as shown in the causal loop diagrams on pages 202, 203 and 204. But if the target setter does not have that authority, then any *incentive* has to work more subtly. How, for example, can a government, seeking to reduce greenhouse gas emissions, persuade people to insulate their houses, but without making failure to do so a criminal offence?

Recognising that *action takers* must be *motivated* to *do the right thing*, the target setter (the government) needs to determine a suitable *incentive* (in this case, say, a cash grant) so that *action takers* (the public) can realise their *desire for something they want* (more cash) only by taking the *right action* (to install insulation).

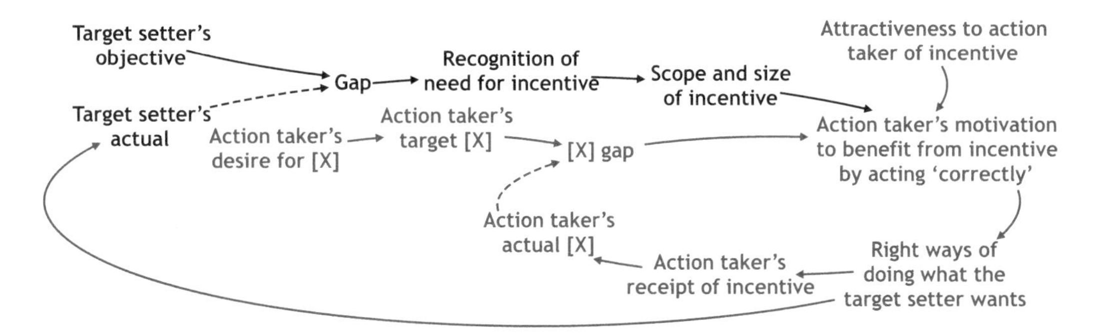

This system comprises two linked 'big is good' (in this example) balancing loops, one for the target setter, the other for the action taker, such that the '*right actions*' are the same for both loops. For this to happen, the *incentive* must satisfy the *action taker's desire*, and be *sufficiently attractive* to drive the appropriate *action*. Designing incentives that motivate people to make the 'right' choices, whilst disincentivising wrong ones, often requires very careful thought – which is where systems thinking, and drawing causal loop diagrams like this, can really help, as discussed on pages 310 and 311. Also, this theme is central to 'Behavioural Economics' in general, and to 'nudge' theory in particular, as advocated by Richard Thaler (who won the 2017 Nobel Prize in Economics) and Cass Sunstein in their best-selling book *Nudge – Improving decisions about health, wealth and happiness*, a title that exemplifies the 'nudge' philosophy, for its 'promise' surely incentivises a prospective reader to buy the book!

Chapter 14

Teachers Behaving Badly

DOI: 10.4324/9781003304050-16

Teachers Behaving Badly

> ***An investigation into the true scale of 'off-rolling' from schools in England has found that more than 49,000 pupils from a single cohort disappeared from the school rolls without explanation.***
>
> ***Researchers from the Education Policy Institute (EPI) said that one in 12 pupils (8.1%) from the national cohort who began secondary school in 2012 and finished in 2017 were removed from rolls at some point, for unknown reasons.***
>
> ***Off-rolling is the practice whereby schools remove difficult or low-achieving pupils from their rolls so that they are not included in their GCSE results, or in order to reduce costs.***
>
> ***The Guardian, 18 April 2019***
>
> *https://www.theguardian.com/education/2019/apr/18/more-than-49000-pupils-disappeared-from-schools-study.*

That's an extract from a report in the newspaper *The Guardian* on 18 April 2019.

A brief explanation for those unfamiliar with the school examination system in England… 'GCSE' is a public examination taken by students aged about 16, and 'A level' at about 18. Both exams are very important, for the grades awarded matter. Many universities take GCSE and A level grades into consideration in determining which students are to be admitted to particular programmes, and many employers have recruitment requirements such as 'must possess a minimum of [this GCSE grade] in [this subject]'.

For a student to be denied the opportunity to even sit the examination is therefore potentially life-changing. Yet, as this report states, some 8% of the total number of students who should have taken GCSE exams didn't. And one reason is so that the low grades likely to be awarded to weaker students are 'not included in the school's GCSE results' – and so do not 'damage' the school's performance measures…

Teachers Are Under Pressure...

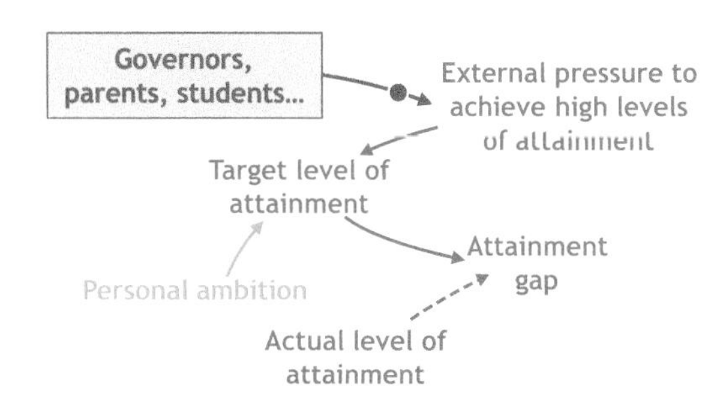

Schools, and their Head Teachers, have a number of performance measures, one of the most important being to achieve one or more *target levels of attainment*, such that the higher the *attainment*, the better – so 'big really is good' in this context.

Given the very many activities carried out within a school, there are many possible *target levels of attainment* that could be of relevance, for example, measures relating to sport (number of students participating in school teams, number of students selected to play for England...), drama (number of school plays produced...), community involvement (number of local support projects...), student well-being (survey of student 'happiness'...) and of course exam results (percentage of students awarded top grades).

Whatever the *target levels of attainment* might be for any particular school, the Head Teacher is under *external pressure to meet them*, certainly *pressure* from, for example, the school's *Governors*, and quite likely from *parents* and *students* too, for all these stakeholders wish to be associated with a 'good' school.

Furthermore, the Head Teacher him- or herself will want to do a 'good job', and to be seen to be doing a 'good job', and so is self-motivated by personal pride and *personal ambition*.

It could be the that school is already meeting its *target levels of attainment* quite comfortably, in which case the Head Teacher just needs to maintain things as they are, or indeed might be motivated to exceed current performance.

Alternatively, one or more of the *target levels of attainment* might not be being achieved, in which case there is a corresponding *attainment gap* between the *target* and the corresponding current *actual*.

Some Important Considerations

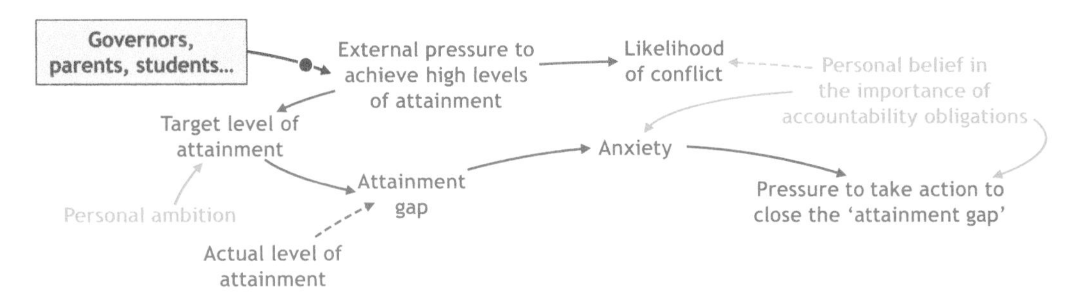

If the Head Teacher believes that the *accountability obligations*, and the corresponding performance measures and *target levels of attainment*, are valid, then the Head Teacher will be highly *anxious* to achieve them, and will wish to *take action* accordingly.

But it could be that the Head Teacher believes that some, or all, of the *accountability obligations* are wrong – for example, a Head Teacher who genuinely believes in the importance of developing the 'whole person' is likely to oppose *accountability obligations* that focus solely on examination results. A low level of *personal belief in the importance of the current accountability obligations* will, in the first instance, lead to a *reluctance to take any particular actions to close the gap.*

This *reluctance* will quickly be noticed by the authorities that wish to impose the *accountability obligations*, resulting in an increasing *likelihood of conflict*. Any *conflict* has just one of three possible outcomes: the Head Teacher concedes, and 'follows the rules'; or agreement is reached to *review, and possibly modify*, the performance measures and *target levels of attainment*; or the Head Teacher '*retires*'. These three possibilities are not explicitly shown in this causal loop diagram at the top of the page, but they could be by introducing further variables and links, as, for example, suggested in the fragment below.

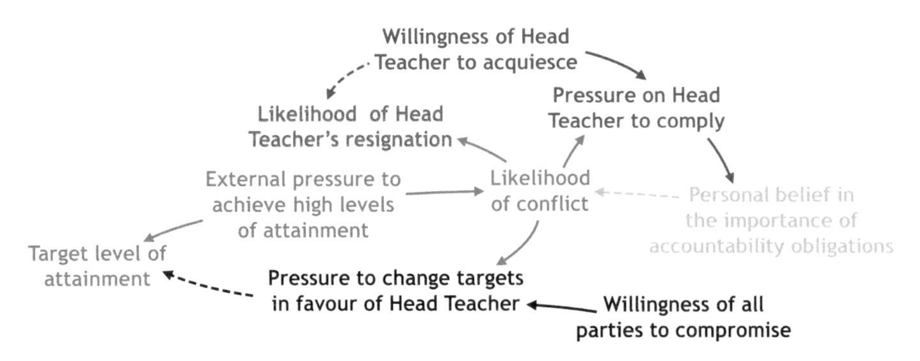

This is an example of an important feature of all causal loop diagrams: they are never 'complete', for something is always left out. What's 'in' and what's 'out' is a matter of judgement, often with a view to avoiding 'clutter'. Such judgements, however, should always be kept under review…

How Important Are Exams?

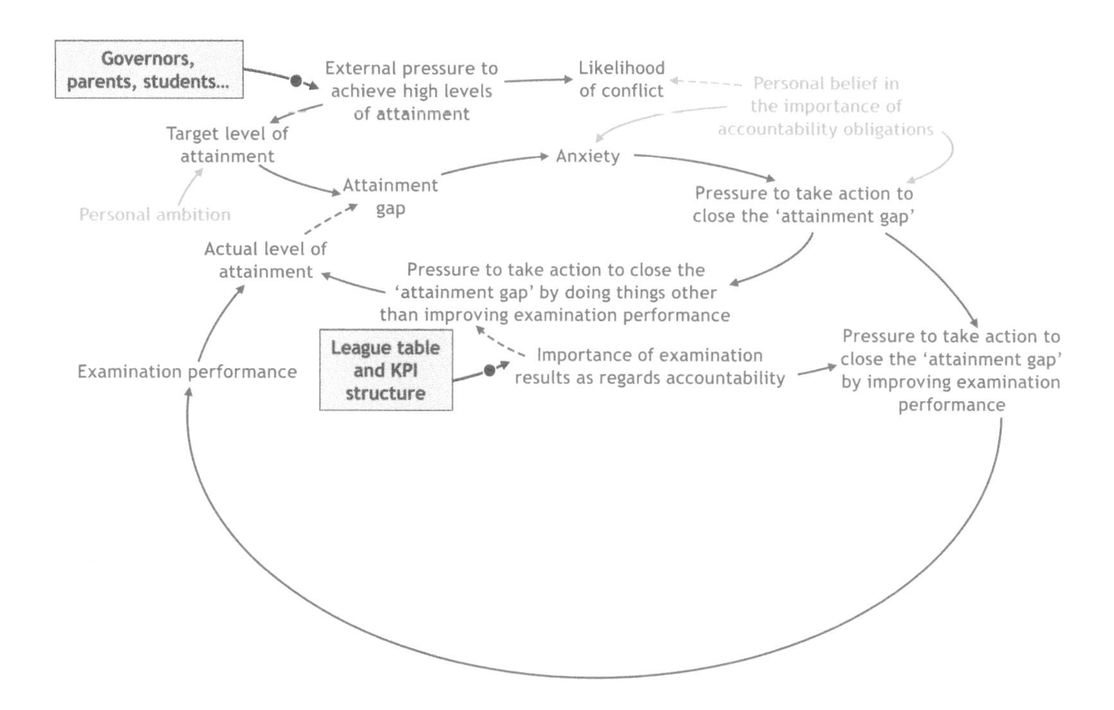

This captures the idea that the *actions that need to be taken to close the attainment gap* depend on how the required *attainment* is specified – with particular reference to the relative importance, or otherwise, of *examination performance*. The identification of what is important is made explicit by the definition of the *key performance indicators* (*KPIs*) by which the Head Teacher will be measured, and in the structure of any League Tables that might be published. For schools in England, for example, the government publishes League Tables that show five measures of examination performance, including the percentage of pupils achieving good grades in GCSE English and Maths – see, for example, https://www.compare-school-performance.service.gov.uk/schools-by-type?step=default&table=schools®ion=all-england&for=secondary&basedon=Overall performance&show=All pupils.

'Learning' Cultures

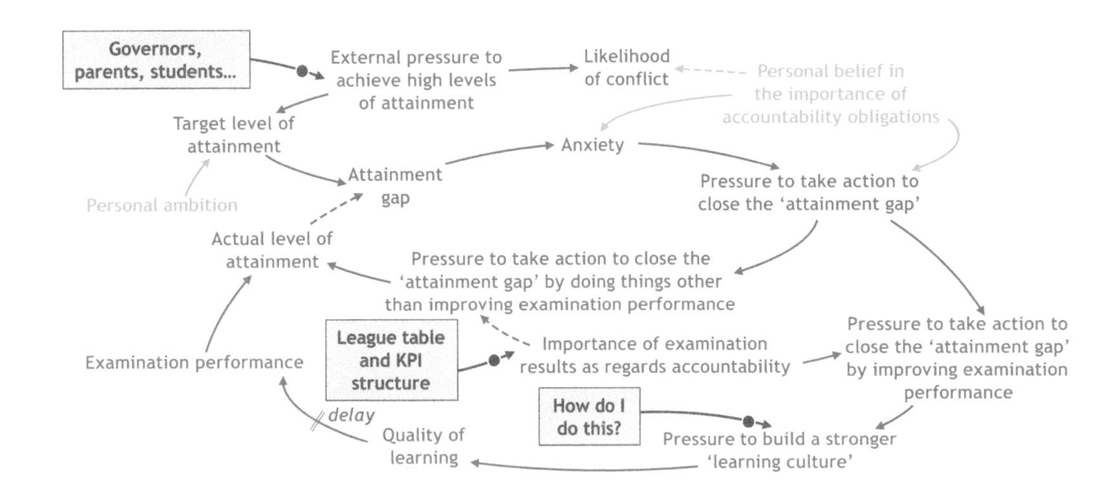

Suppose a Head Teacher wishes to *improve the school's examination performance*, perhaps substantially.

What does the Head Teacher *actually do*? The Head Teacher reads some books, takes some advice, attends some conferences. The key to success, it seems, is *to build a stronger 'learning culture'*, in which students are eager to learn, and teachers are inspiring.

That's great!

Just two problems.

What does the Head Teacher *actually do*, tomorrow, next week, next month… to *build a 'learning culture'*? It all sounds very appealing, but it's a bit vague… and perhaps the Head Teacher does not feel confident in *actually doing* it…

And how *long will it all take* before improved examination results come through? The *Board of Governors* are pressing for an improvement next year…

...and 'Better' Teachers

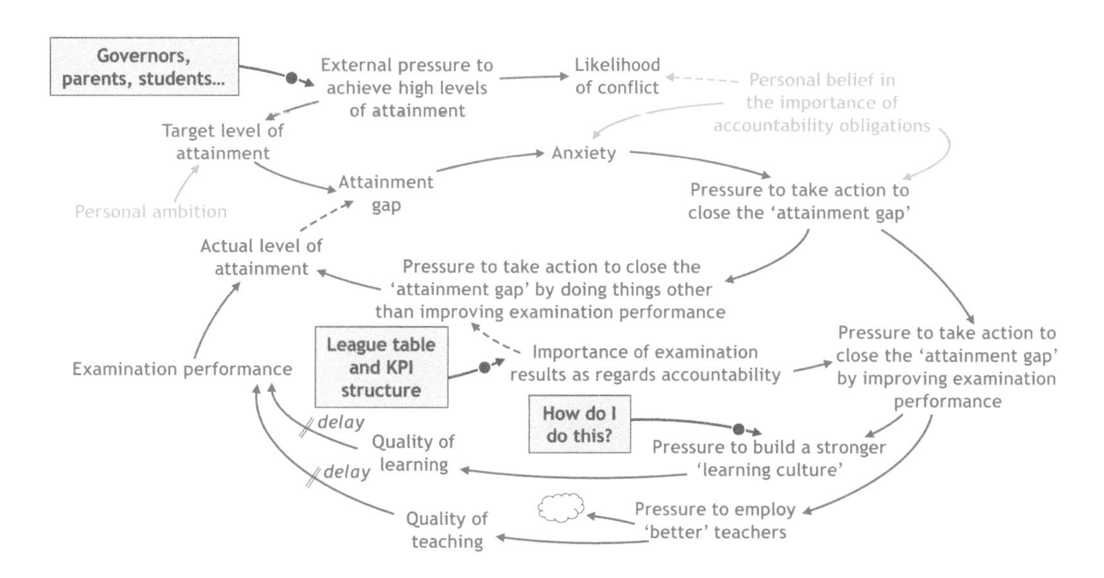

Or maybe the 'answer' is to *employ 'better' teachers* – teachers who are more energetic, innovative, charismatic...

That's not so easy, either. Where are these *'better' teachers*? What would attract them to this school? What if a newly appointed teacher turns out to be not-so-good? And – once again – *how long* before better examination results come through?

But perhaps the biggest problem might not be the recruitment of new teachers, but how to 'exit' the existing ones – especially if the Head Teacher is, by temperament, someone who seeks to avoid confrontation. That can be very messy... as indicated in the causal loop diagram by the 'cloud', which signals that there are further consequences that are not detailed here.

'Taking Advantage'

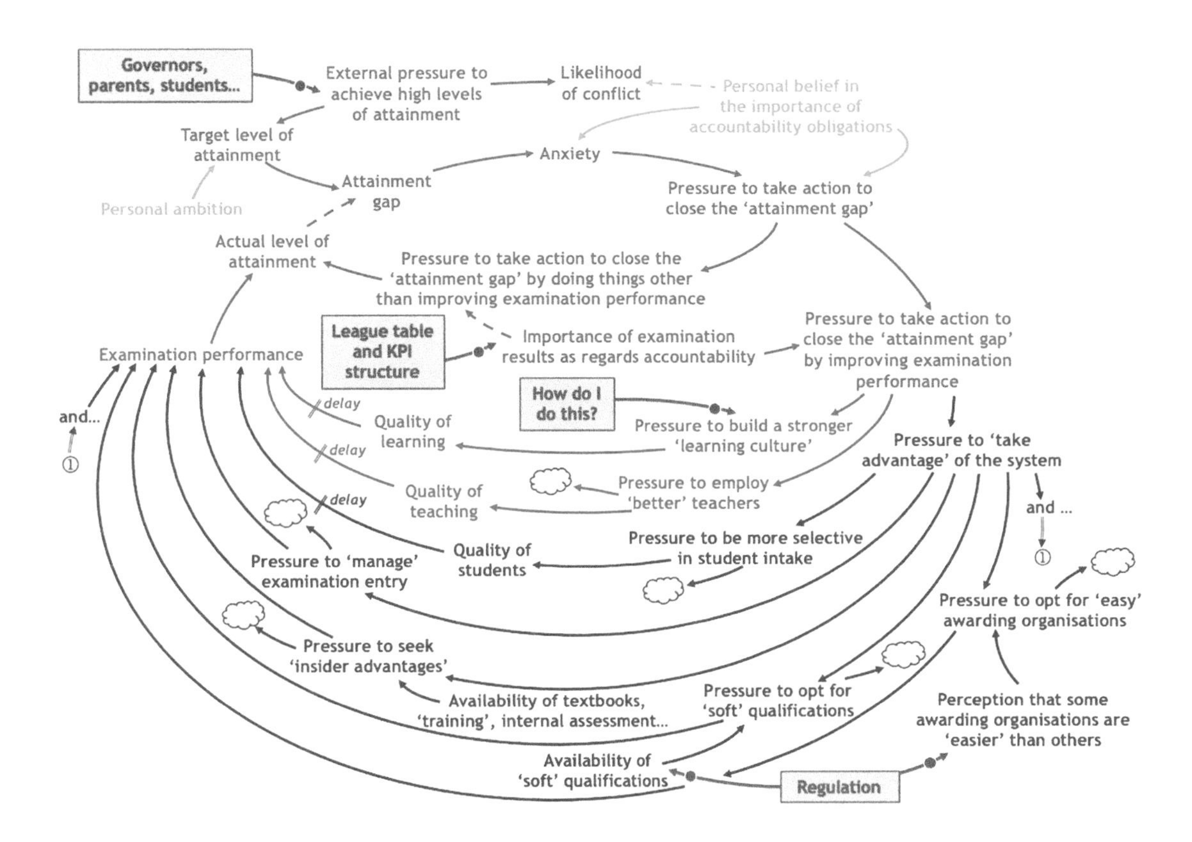

For narrative, see following page

'Taking Advantage'

And, without any doubt, building a 'learning culture' and re-configuring the teaching staff are both hard things to do.

Are there any easier ways? Ways that deliver better *examination performance* but much more quickly and without so much hard work? Yes, there are... for example...

If only more able students are admitted to a school, then there is a higher likelihood that they will achieve better *examination results*. Legally, in England, many – but not all – schools are not allowed to *'cherry pick' their students*. But some can...

But even if a school cannot be selective in the students admitted, perhaps it can *'manage' who does, and who does not, sit any particular exam*. Perhaps weaker students can be 'encouraged' to drop that subject, or even 'persuaded' to leave the school – this being 'off-rolling' as reported in the newspaper story on page 208.

In England, examinations are set and marked by three different, competing, 'awarding organisations', and schools can choose to use exams set and marked by whichever of these organisations they wish. If an organisation publishes any textbooks, or other relevant material, in a particular subject (as is actually the case), might those indicate the likely content of any exam? And if a teacher is helping an awarding organisation in setting this year's exam, might that 'insider knowledge' be 'useful' when giving lessons, or setting tests?

In which case, is it in the school's interests to have as many teachers as possible working with the awarding organisations?

Physics is a notoriously 'hard' subject. Rather than encouraging a student to study Physics, and risking a lower grade, should that student be 'persuaded' to study a 'softer' subject, and so be more likely to get a higher grade? This may be in the interests of the school, but is 'managing' the subjects a student is allowed to study in the best interests of the student?

And does one particular 'awarding organisation' set 'easier' exams in any given subject than the others? In which case, the school's position in the *League Table* will be higher if the students sit that organisation's exams...

All these are 'easier' ways of closing the *attainment gap*. All are real, and have actually happened. None are 'unintended consequences'; rather, many are the result of a failure to think the system through at the outset, and to close the loopholes...

...as is the case for the two *'soft'* options, for one of the duties of Ofqual, the English *regulator* of school exams, is to ensure that there is a 'level playing field' across all awarding organisations, applying the same standards across all exam subjects.

Justifications?

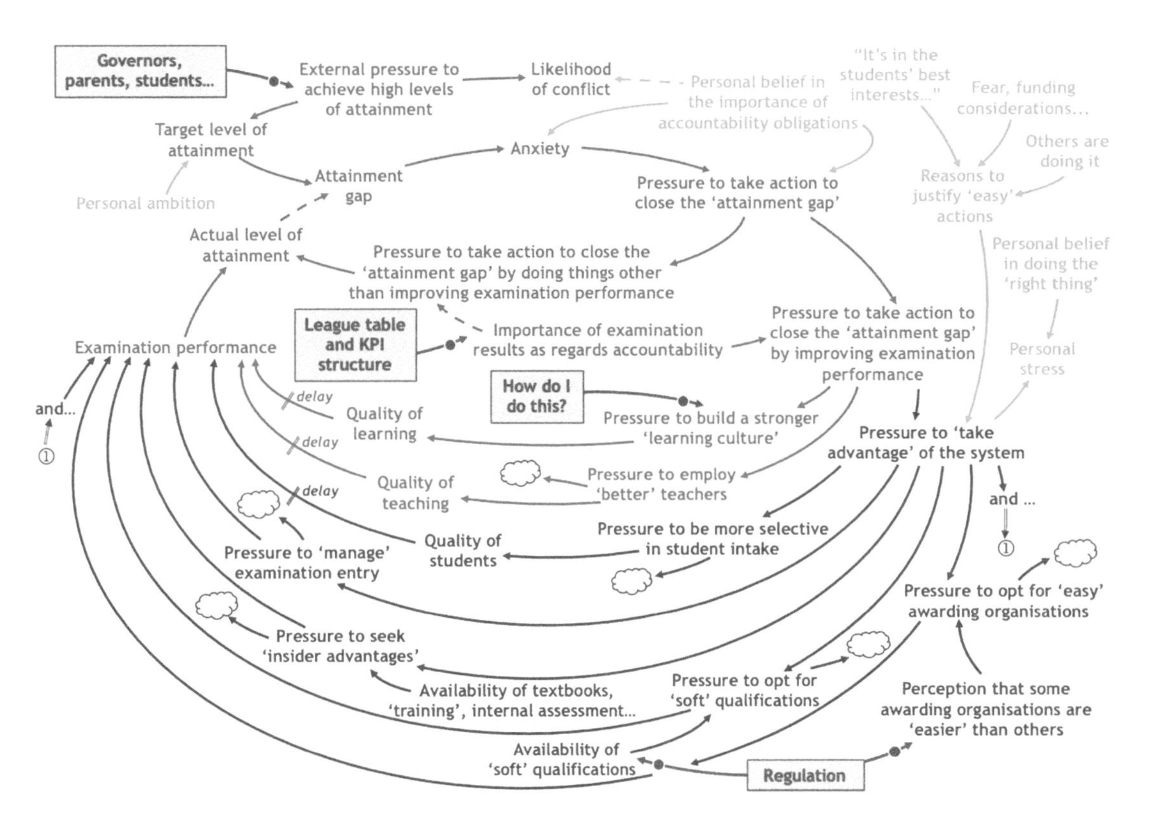

For narrative, see following page

Justifications?

Knowingly choosing an 'easy' way of doing things rather than a 'hard' way can often set up an ethical dilemma, especially if the 'easy' way is acting in the interests of closing the *attainment gap* but not to the benefit of the student.

So the new items on the causal loop diagram on page 216, shown in green on the upper right, identify some of the reasons that a Head Teacher can use to *justify taking the 'easy' actions*. Reasons such as...

All students, naturally, want high grades. So isn't *in the student's best interests* to avoid being given a poor grade by *not sitting the exam at all*? Or by taking *'softer' subject*?

And if the school fails to close the *attainment gap*, that could bring all sorts of trouble – the *Governors* might start thinking of finding a new Head Teacher, and the English regulator of schools, Ofsted (not be confused with the regulator of school exams, a different body called *Ofqual*), might even put the school in what is known by the somewhat Orwellian term 'special measures'...

Furthermore, if *other schools* are 'playing games', and getting away with them, well...

These are all very understandable, and all very real. But the tension between knowingly *'taking advantage' of the system* and *the right thing to do* can cause (considerable) *personal stress...*

Chapter 15

Perverse Incentives and Unintended Consequences

DOI: 10.4324/9781003304050-17

Oh, No! That's Not What I Wanted to Happen…

An unintended consequence (or perverse incentive) is a driver that works against the objectives sought from the instrument. Perverse incentives produce unintended consequences, which may provide unexpected benefits or costs to business or government. For example, a law providing a reward for control of a pest may encourage individuals to farm the pest to claim the reward.

From the website of the New Zealand Ministry for the Environment

https://environment.govt.nz/publications/waste-policy-discussion-the-potential-unintended-consequences-of-a-national-waste-levy/4-definition-of-unintended-consequences/.

The New Zealand government did not make that example up. It actually happened in Hanoi in 1902. With the intention of ridding the city of rats, the authorities offered a bounty to kill them, payable on the presentation of each dead rat's tail. Two things then happened. Many rats were discovered running around the city, but lacking a tail. And some entrepreneurs discovered that breeding rats had become a very profitable proposition…

Hanoi's Rats

As discussed on pages 199 and 206, *incentives* are often used to encourage the *'right' actions* to close a given *gap*, and *disincentives* to discourage the *'wrong'* ones. Sometimes, however, this can go badly awry, and an incentive intended to encourage a 'good' action can turn out to be hugely counter-productive.

An example is the story of Hanoi's rats. As mentioned on the previous page, the government wished to reduce the number of rats in the city. To achieve this, the authorities might have hired some rat-catchers, but instead decided to deploy the local population. *Recognising* the need to *motivate* Hanoi's citizens to catch rats, the authorities offered the incentive of a *bounty on rat tails*, as shown in the causal loop diagram here, which is based on the generic target setter – action taker diagram shown on page 206, where 'small is good' as regards the rats, and 'big is good' as regards the action taker's cash.

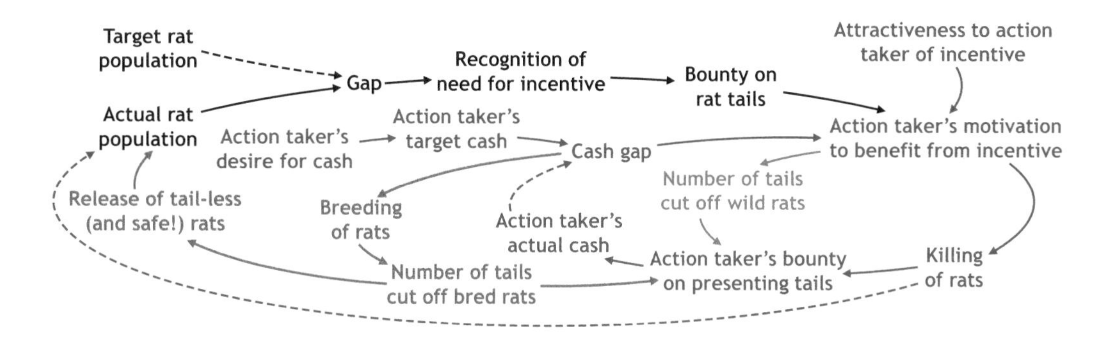

The government's intention was that the *bounty* would encourage the *killing of rats,* decreasing the *actual rat population* to the (very low) *target.* What the government didn't spot was that the action taker's balancing loop could be closed by claiming the *bounty on presenting a rat tail*, but without *killing the rat.* Nor that *breeding rats* would readily meet the *action taker's desire for cash*, whilst making matters worse by increasing the *actual rat population* as the result of *releasing* all those *specially bred rats*, once their *tails had been cut off* and the *bounty claimed* – rats likely to survive as they no longer 'qualified' for the bounty!

The incentive of the *bounty on rat tails* resulted in outcomes quite different from those intended by the government, this being just one instance of a **perverse incentive** driving **unintended consequences**. There are many more…

Some Unintended Consequences of Biomass Fuel

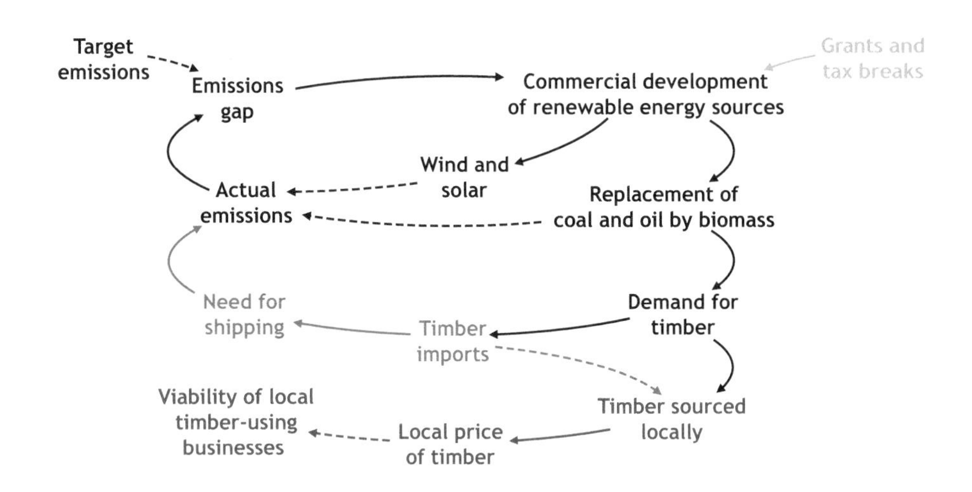

Since 2005, the price of the kind of wood used in the construction and wood panel industries has gone up by more than 50% in the UK. The reason, the wood industry says, could not be simpler – biomass subsidies have increased demand for wood, pushing prices higher. The effects are already being felt.

'The increases in costs eroded our bottom line significantly, until the point when, about 12 months ago, we were forced to pass them on to our customers', says Gavin Adkins at wood panel manufacturer Kronospan. The customers of panel makers are furniture makers, and they too are feeling the heat.

BBC News, 28 February 2012

https://www.bbc.co.uk/news/business-15756074.

Unintended Consequences

As the (true) story of the teachers, described on page 208, shows, the performance measure requiring schools to deliver high exam grades did not result, as intended, in driving up standards of teaching and learning across the country; rather, it triggered a variety of 'easy' ways by which some schools cheated the system. These are examples of so-called 'unintended consequences' – outcomes which actually happen, but which were not the original objective.

Sometimes, the unintended consequence is 'a good thing', an example being the re-wilding of the 'no-man's land' at the border between North and South Korea. When each side built their barbed-wire fences to prevent access, they were not doing that with the intention of creating a human-free, and therefore thriving, natural wildlife habitat. But that's what happened.

'Good' unintended consequences are rare; 'bad' ones are much more frequent. Many are associated with easy (= bad) ways of achieving performance measures, targets, objectives or goals, as illustrated by the story of the teachers; sometimes the unintended consequences are the result of an unwillingness to accept the target in the first place; sometimes the objective is achieved, but itself triggers bad things to happen (between 1920 and 1933 in the US, the production and sale of alcohol was banned, driving the trade underground and causing an increase in crime and corruption); sometimes the unintended consequences result from the (intended) action, as illustrated by the example of biomass fuels, as shown in the causal loop diagram on page 222.

The context is a government that wishes to reduce emissions of carbon dioxide, and sets a low *target* accordingly. To implement that policy, the government seeks to encourage the *commercial development of renewable energy sources*, such as *wind and solar*, and the *replacement of coal and oil by biomass*, such as wood pellets (although biomass is burnt, it is in principle renewable, for the trees from which biomass is made can be replaced by new planting).

The installation of biomass-fuelled power stations increases the *demand for timber*, and timber-derived products such as chippings. Trees grow slowly, and in any locality, supply is limited, so if that *demand* is met *locally*, the *local timber price* will rise, increasing the costs of local businesses that use timber, such as furniture makers, or manufacturers of garden sheds. This could *put those businesses in commercial jeopardy*, causing them to lobby the government Minister for Rural Economies, and sparking a political row with the Minister for Energy about not having thought through the consequences of the biomass policy.

This problem can be resolved by *importing timber* to meet the *biomass demand*, so reducing the *local demand* and (in so far as timber prices are not global) causing the *local price of timber* to fall. But those *timber exports* need to be *shipped in*, and the fuel burnt to do that increases *emissions*. Overall, a policy intended to reduce emissions ends up increasing them. And incentives such as *government grants and tax breaks* make matters even worse. A perverse incentive indeed.

Hoover UK's Free Flight Promotion, 1992

Hoover UK's Free Flight Promotion – The Intent

Perhaps the most spectacular – if not notorious – example of a perverse incentive was Hoover UK's 1992 free flight promotion. Hoover UK wanted to generate additional profits by stimulating sales of their washing machines, vacuum cleaners and other domestic appliances, but they knew, of course, that no manufacturer can 'force' someone to make a purchase. Hoover UK were therefore a target setter, seeking to motivate a potential purchaser, the action taker, to choose Hoover rather than another brand. Accordingly, Hoover UK decided to offer the incentive of two free flights to the US, this being rather a more exciting sales promotion than 'five boxes of soap powder free too!'

This much-simplified version of the diagram shown on page 206 captures what the Hoover UK managers who designed the scheme had in mind:

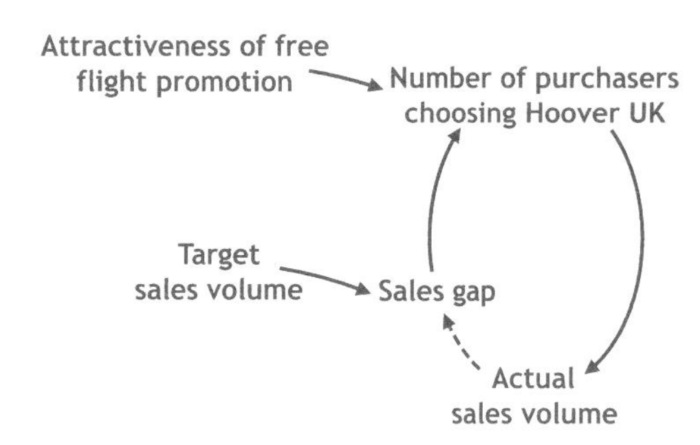

The promotion worked, in the sense that it did indeed stimulate purchasers to choose Hoover UK. But many other things happened too, all of which were undoubtedly unintended…

Hoover UK's Free Flight Promotion – What Actually Happened

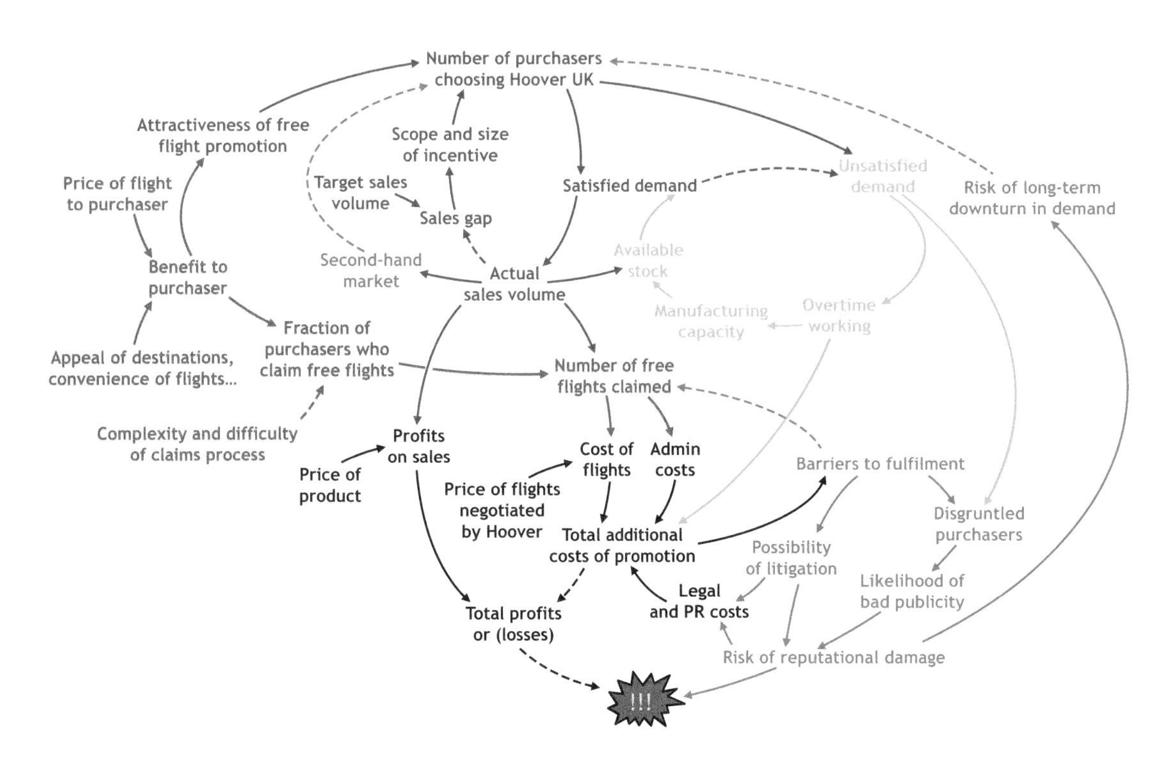

This causal loop diagram captures the key features of what actually happened.

In 1992, Hoover UK wanted to increase their *sales volume*. To achieve this, they *recognised the need to encourage* more potential *purchasers to choose to buy a Hoover UK product rather than a competitor's*. They therefore decided to offer the incentive of two *free flights* to the US with all purchases over £100. This was widely advertised, and proved to be very *attractive*, for it gave the *purchaser a considerable benefit*, especially for *highly-priced* and *convenient flights* to *appealing destinations*. Fantastic! Increased sales. Just what Hoover wanted!

The 'Free Flight Fiasco'

But the good news faded fast. The surge in *sales* quickly depleted the *available stocks* of Hoover UK products, exceeding the rate at which the *manufacturing capacity* could replenish supply. As discussed on pages 101, 109 and 114, *satisfied demand* is the lesser of the *number of purchasers who want to buy a Hoover product* and the *available stock*, and any further *demand remains unsatisfied*. This had two consequences: purchasers who were unable to buy a product had to forego not just the product but also the two flights, and so were *disgruntled*; also, the *unsatisfied demand* acted as a trigger to increase *manufacturing capacity* by *working overtime*. This increased *costs* and eroded *profits*, which was of course totally counter-productive, given that the original objective was to increase *profits* by boosting the *sales volume*.

Furthermore, when people realised that the cost of two tickets to the US was more than the £100 minimum spend on a domestic appliance, that served as an incentive to buy a product they did not want, and so, after a short delay, the *second-hand market* in nearly new appliances became very active, so depleting the *demand for new products* from Hoover UK. Oh dear.

As *product sales* boomed, so did the *number of free flights claimed*. And the corresponding *costs*. In an attempt to limit the damage, Hoover took various measures to make the *process for fulfilling claims even more difficult* than it was to *make the claim in the first place* – for example, by offering flights from airports deliberately chosen to be a long way from a claimant's home in the hope that the customer would turn the offer down. So Hoover were simultaneously running an incentive scheme to sell their products, and a disincentive scheme to deter purchasers from benefitting from the original incentive!

This created a huge backlash. Frustrated claimants were sorely *disgruntled*, so much so that in 1993 some formed a pressure group, taking Hoover to *court*. And the 'Free Flight Fiasco' story was *covered in the press and on television*.

An incentive intended to increase Hoover UK's *profits* actually resulted in (1) losses estimated at around £50 million; (2) the firing, in March 1993, of the three senior executives responsible for the scheme; (3) the sale, in 1995, of the Hoover business in Europe at a loss of $81 million to the American parent, Maytag; (4) a sequence of law suits that dragged on until 1998.

The *free flight promotion* was indeed a perverse incentive. And the resulting wound was totally self-inflicted, for all of the outcomes could have been anticipated had the Hoover UK management team just thought things through rather harder. If only they had compiled a causal loop diagram like the one on page 226 before they took the decision...

A Reflection on Objectives and Performance Measures

As the (real) stories of 'ambulance stacking', 'teachers behaving badly', biomass, Hanoi's rats, and Hoover UK's free flight promotion make very clear, introducing targets, goals, objectives and performance measures in the hope that the 'simple' balancing loop will result in 'all being well' is, in any real system involving real human beings, likely to be somewhat naïve.

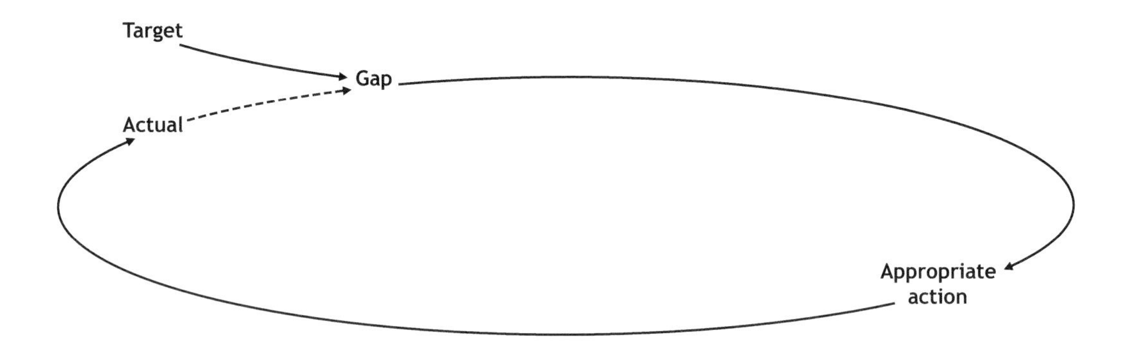

The Truth About Balancing Loops

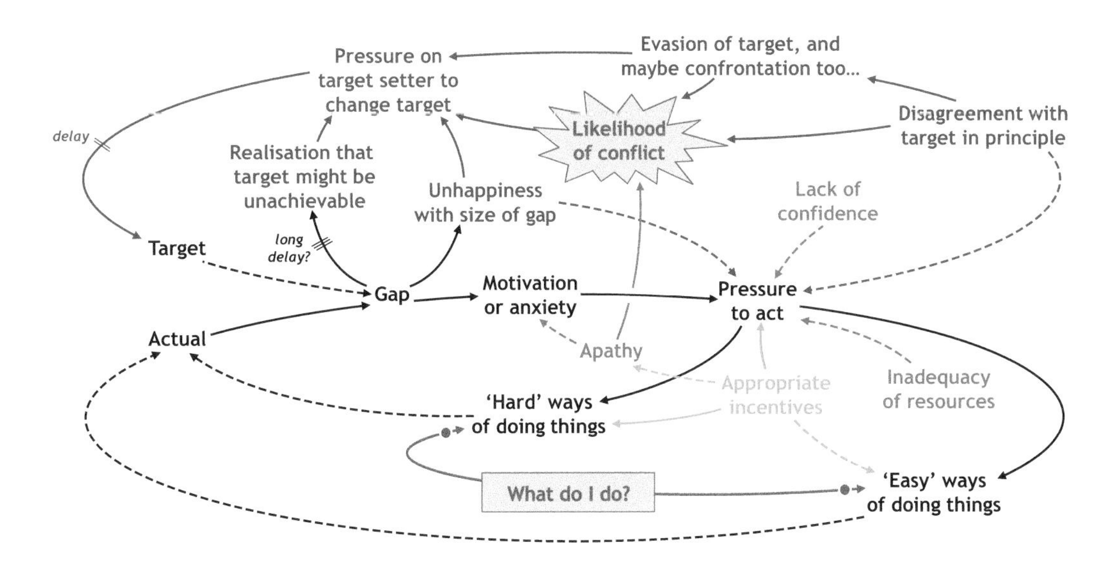

Far wiser to start with this much more realistic causal loop diagram, so anticipating, for example, *apathy*, and the likelihood of there being *'easy', but wrong, ways of doing things,* as well as *'hard', but right, ways.*

Also, remember the difference between target setters and action takers (as discussed on page 206), beware perverse incentives, and always bear in mind that there are never any unintended consequences. But there is much evidence of poor systems design, and even poorer thinking.

The next page therefore offers some practical advice as to how to design effective systems, avoiding the twin dangers of perverse incentives and unintended consequences.

How to Avoid These Twin Dangers

Whenever there are performance measures, objectives, goals and targets, whenever there is a policy to deliver an election promise, there will always be one or more corresponding balancing loops that specify whatever action is required to transform the current state into the desired state, as specified by that target or policy objective. And wherever there is a balancing loop, there is always the possibility of unintended consequences. And whenever an incentive is offered to encourage the 'right' action, there is always the possibility that any incentive might turn out to be perverse.

So in thinking about performance measures, objectives, goals, targets and policies, and before the associated procedures and systems are designed and implemented, future problems can be avoided by anticipating, as comprehensively as you can, what those unintended consequences might be, and how any well-intentioned incentives might turn out to be perverse. And when you've done that, you can design them all out. Here are some suggestions as to how this can be done...

1: Draw as comprehensive a causal loop diagram, of the whole system, as you can.

2: Identify as many possible actions that might be taken to close any 'gap' as you can. Which are 'right'? Which are 'wrong'?

3: Given the target, how might the most unscrupulous person cheat to achieve it?

4: If someone thinks the gap to be closed is too big, or disagrees with the target in principle, what might happen?

5: How might 'right' actions be incentivised? And 'wrong' ones discouraged?

6: If the target setter has little authority, how many 'desires' can you identify that might act as suitable incentives to increase the motivation of the action takers? How might each of these different 'desires' be satisfied?

7: For each incentive, and disincentive, what actions might these also encourage?

8: The purpose of any incentive is to encourage a particular action. The overall objective, however, is rarely to encourage an action; rather, it's to achieve an outcome. The action is a means-to-an-end, but action-for-its-own-sake is not what's wanted – what's wanted is the right outcome. What needs to be done to ensure that the right outcome is achieved, rather than lots of action?

9: Beware group-think. In addressing these questions, people who see the world as you do, and who think like you do, will come to the same conclusions. But unintended consequences arise, and incentives become perverse, because those who are tasked to close the gap aren't like you. They see the world differently. How can you gain insight into those different world-views, before it's too late?

Chapter 16

Delivering General Practice

DOI: 10.4324/9781003304050-18

GP Practices at Breaking Point

Doctors are warning that general practice clinics risk cracking under the pressure of 'unsustainable' workloads unless the government ramps up the recruitment of medical staff and takes steps to reduce burnout.

The Royal College of General Practitioners is calling on the government to introduce an emergency rescue package to shore up general practice clinics after the pandemic, including recruiting 6,000 more GPs and 26,000 additional support staff, such as nurses and receptionists, by 2024 as well as reducing paperwork and investing in £1bn worth of improvements to infrastructure and technology. Without these changes, patients will not receive the care they need, the college said.

The Guardian, 29 July 2021

https://www.theguardian.com/society/2021/jul/29/gp-clinics-at-breaking-point-and-recovery-plan-is-essential.

The Covid-19 pandemic has stretched healthcare services around the globe to the limit. In England, where this story is set, the front line of medical care is – as it is in many other countries too – the local family doctor, the general practitioner, the GP....

A GP Practice's Delivery Depends on Capacity and Capability...

In many countries, anyone in need of emergency medical help – for example, after an accident – will be taken directly to a hospital. For non-emergency conditions, however, the patient goes to the local General Practitioner's surgery to take a seat in the waiting room, perhaps having previously made an appointment. The GP practice has no control over the demand for its services: patients arrive as they need to, so determining the *number of cases presented to the GP practice per day*.

Every GP wants the outcome of each consultation to be the best possible for each patient. Ideally, each patient is restored to normal health; if that is not possible, then the condition should satisfactorily be brought under control. This concept is represented in the diagram by the *number of patients successfully managed per day*. As will be discussed on page 237, the qualifier 'successfully managed' is important, for it recognises the possibility that some conditions, as presented, might not be fully treated by the GP, but require access to facilities beyond the GP practice (for example, an X-ray machine), referral to an appropriate medical specialist, or admission to a hospital.

One key determinant of *the number of patients successful managed per day* is the *practice total capacity and capability* – the *bigger the practice*, the greater the *number of patients* that can be seen over any time period; the *broader the capability*, the greater the range of conditions that can be treated without the need to refer the patient elsewhere. Since the *number of patients successfully managed per day* cannot exceed the *practice's capacity and capability*, the *number of patients successfully managed* may, in the first instance, be expressed as a MINIMUM function of the form

$$\text{MIN}(\textit{Cases presented, total capacity and capability})$$

This is similar to hospital admissions, as discussed on pages 110 and 111.

...Which Depends on Funding and the Local Demographics

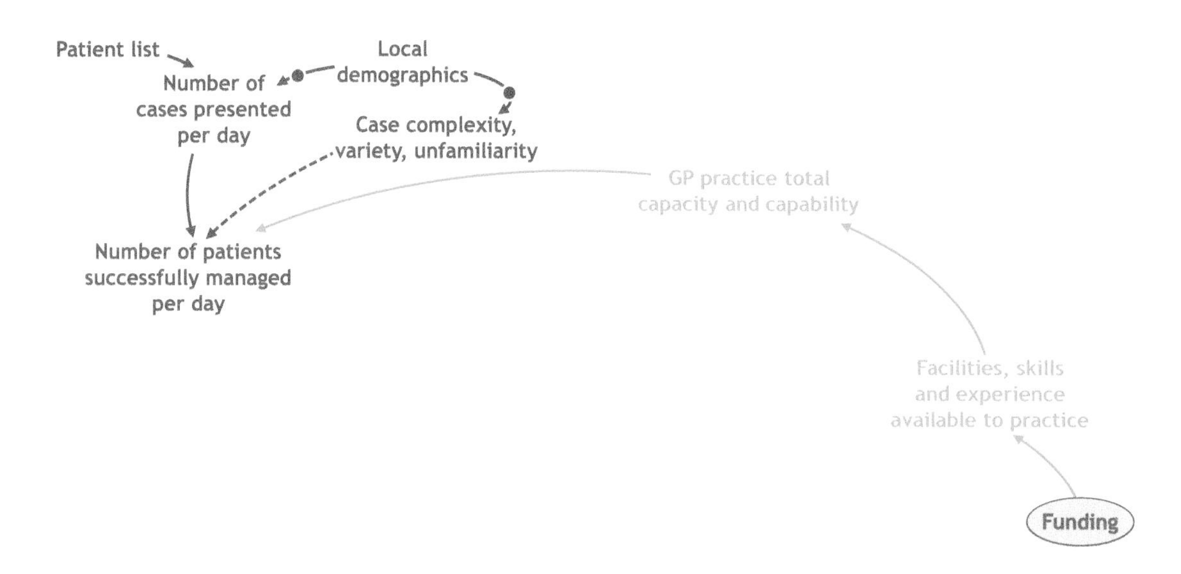

In England, the local population is assigned to GP practices according to each practice's *patient list*. In general, the greater the number of people on a practice's *list*, the greater the *number of cases presented per day*. However, for any given *practice capacity and capability*, the greater the *complexity, variety and unfamiliarity of the cases presented*, the smaller the *number of patients successfully managed*. That's because more time might be required for each patient so fewer patients are seen in any one day; alternatively, if the condition requires referral elsewhere, then although the patient is seen, the condition is not *successfully managed*, hence the inverse link.

The *local demographics* are important too, influencing both the *number of cases presented* and also the *complexity* – for example, elderly people are likely to require a GP's services more frequently, and to present more complex conditions, as compared to younger people.

The *practice capacity and capability* is determined by the *practice's facilities, skills and experience*, themselves determined by current and historic *funding*.

All GPs Want to Deliver High-Quality Outcomes…

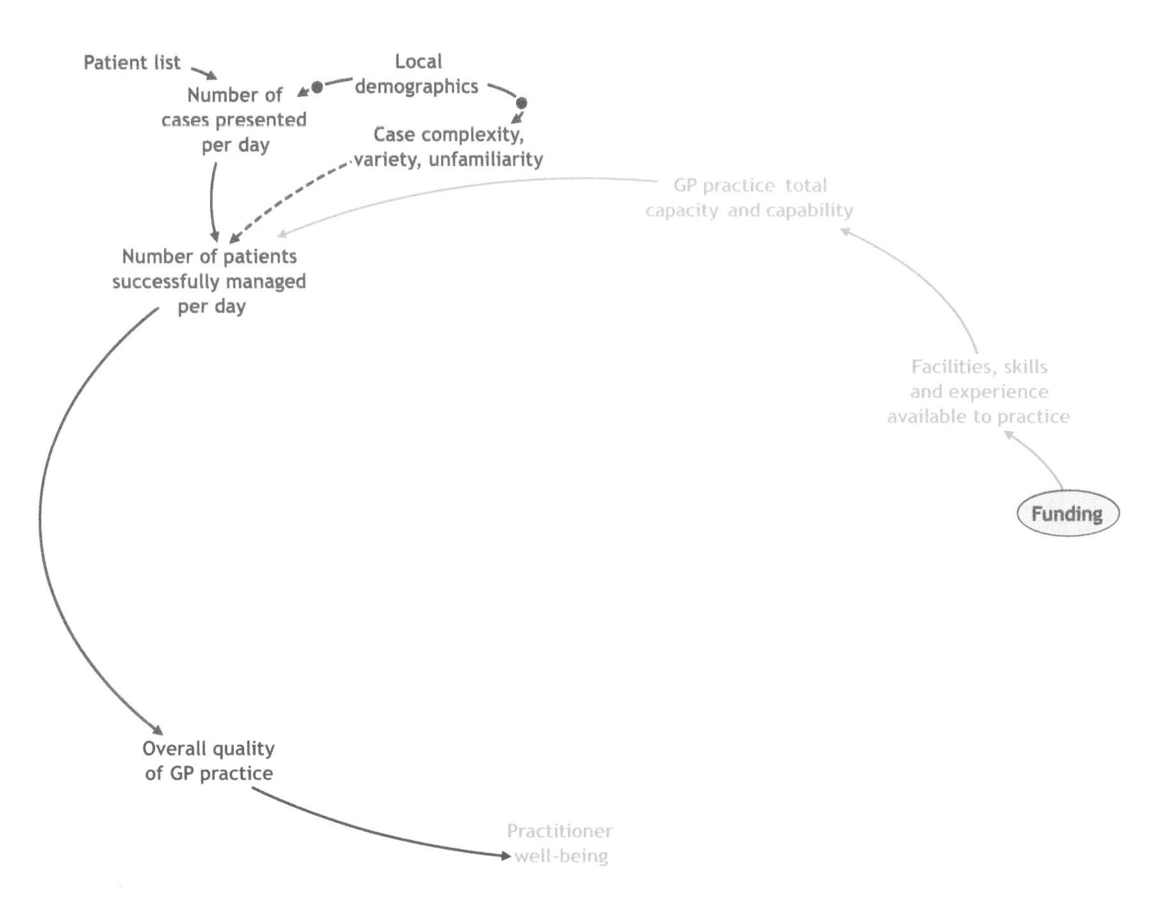

The greater the *number of patients successfully managed per day*, the greater the *overall quality of the practice* within which the GP works. Being part of a high-quality practice, and contributing to that quality, are 'good things', and so the higher the *practice's quality*, the greater the *GP's well-being*.

...but If Patient Demand Exceeds the GP Practice's Capacity...

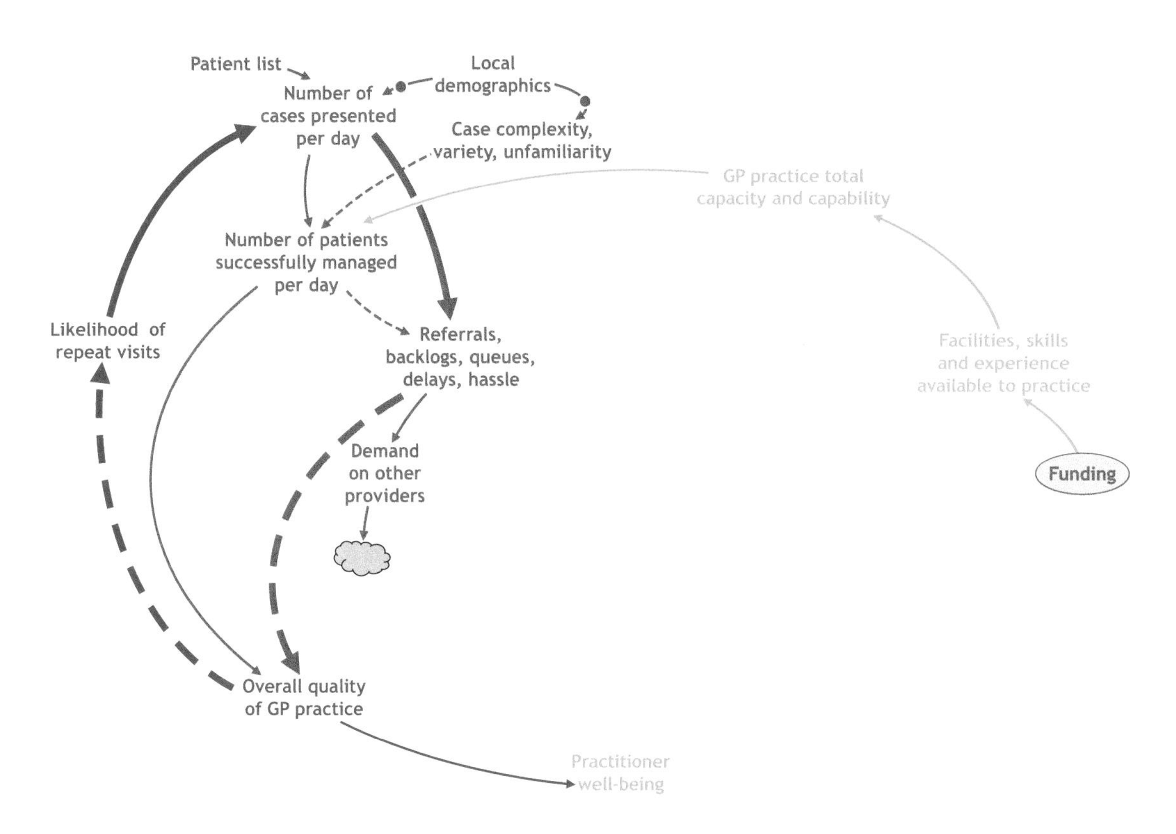

For narrative, see following page

...Backlogs and Queues Emerge, and Get Longer...

If the *number of cases presenting per day* exceeds the *practice capacity and capability*, one of two different things can happen. The first concerns what happens when a patient presents with a condition that the GP practice does not have the *capability* to treat directly and fully – perhaps, for example, some diagnostic tests are required, tests that can be carried out only at a specialist laboratory; perhaps the practice does not have stocks of the appropriate medication and so the GP writes a prescription to be taken to a pharmacist; perhaps the treatment requires admission to a hospital. The result is to *refer the patient elsewhere*, so creating a *demand on another provider* – as will be discussed further on pages 252 and 253.

The second concerns what happens when the practice has the required *capability*, but lacks the required *capacity* – there just isn't enough time to see all the patients in the waiting area. The result is a *backlog* (for example, a patient who had a scheduled appointment in the morning has to wait until after lunch) or a *queue* (all those people in the waiting room, or on the phone trying to book an appointment), and all the associated *delay and hassle*. All of which diminish the 'patient experience', so eroding the *overall quality of the practice*.

But not just that – if, for whatever reason, a patient does not receive complete treatment, there is a possibility that the patient will have to *make another appointment*, so adding to the 'normal' *number of cases presented* at some time in the future.

As can be seen, the highlighted closed loop from *cases presented*, to *referrals, backlogs and queues*, then *quality* and *likelihood of repeat visits*, back to *cases presented* contains two inverse links. Two is an even number, and so this is a reinforcing loop – a reinforcing loop that continuously grows. Accordingly, unless something is done to break the loop, the *backlogs* and *queues* get ever longer...

There are also two other closed loops, both balancing loops. One, with three inverse links, is from *cases presented* to *patients successfully managed*, and then through *referrals, backlogs and queues*, *quality, likelihood of repeat visits* and back to *cases presented*; the other, with one inverse link, is from *number of cases presented* to *number of cases successfully managed*, then *quality* and *likelihood of repeat visits*, back to *number of cases presented*.

Both interact with the highlighted reinforcing loop, and can slow it down, for the greater the *number of cases successfully managed per day*, the smaller the number of *referrals*, the shorter the *queues, backlogs and delays*, and the lesser the resulting *hassle*.

So anything that can be done to increase the *number of cases successfully managed per day* will help. And as can be seen from the diagram, that can be achieved if it might be possible to increase *the practice total capacity and capability* – as will be discussed on pages 244 and 245.

…and Things Can Get Even Worse…

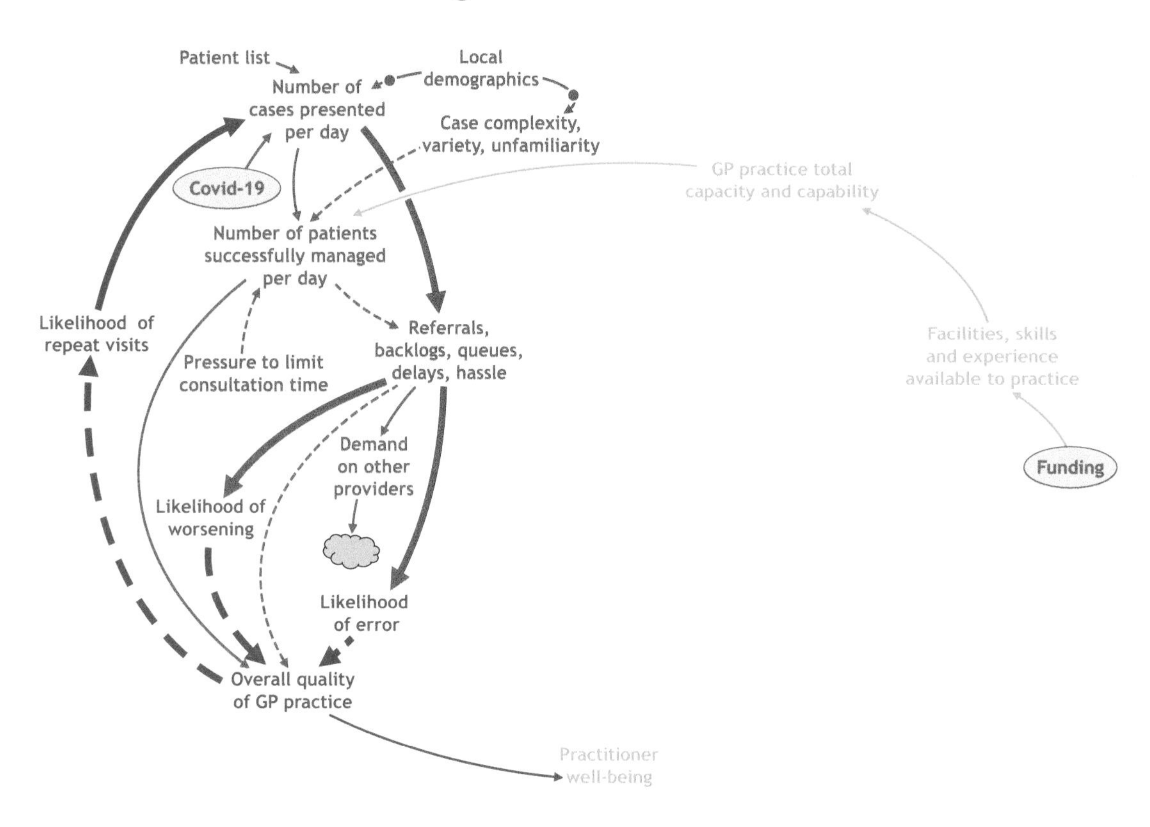

For narrative, see following page

...for Mistakes Can Happen Too...

One of the consequences of *delaying treatment* is the possibility that, whilst the patient is waiting, the *patient's condition can worsen*. And a consequence of that – in addition to the distress caused to the patient – is the likelihood that the patient will need more complex, and more expensive, treatment, so adding even more to the doctor's workload. That sets up a second reinforcing loop, which amplifies the effect of the first one, as discussed on pages 236 and 237.

And there's another consequence too. In the hurry to get things done, the general *hassle* might cause the GP to make a *mistake*, which makes things even worse.

Furthermore, in England over the last several years, there has been increasing pressure on GPs to see as many patients as they can during any day, resulting in a 'standard consultation' lasting 10 minutes – a time which many GPs consider to be too short for many consultations. The stronger the *pressure to limit the consultation time*, the shorter the time allocated to each consultation, and so the greater the *number of patients seen per day* – so that might suggest a direct link between these two variables. That is true. But the *number of patients seen per day* is not the variable shown in the diagram on page 238 – the variable shown is the *number of patients successfully managed per day*, and those words *successfully managed* make all the difference. That's why the link is inverse: the diagram, as shown, suggests that the stronger the *pressure to limit the consultation time*, the shorter that consultation time will be, and so the lower the likelihood that the consultation will result in the successful management of the patient's condition, which, in turn, increases *referrals, backlogs, queues, delays and hassle*, further fuelling those nasty reinforcing loops.

This has been the context in which GPs have been working in England – and quite possibly elsewhere too – for some time.

And then Covid-19 struck. This caused a substantial, sudden, increase in the *number of cases presented per day* at every GP practice across the land. Even those that were just-about coping beforehand were now faced with the necessity of rationing how to spend their necessarily limited time, and of being forced to take very difficult decisions of prioritising one patient over another, resulting in *referrals, backlogs, delays and hassle*, and all the corresponding consequences.

One Way of Dealing with the Backlog…

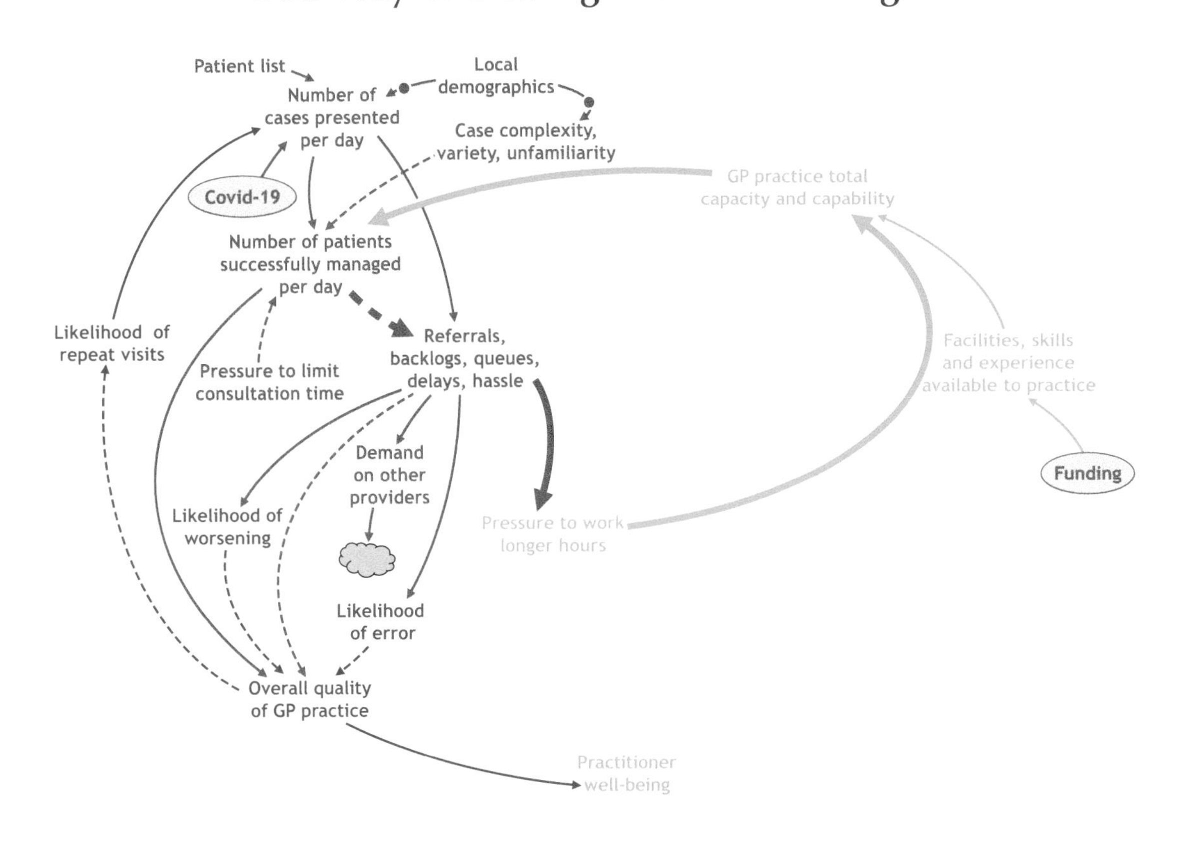

For narrative, see following page

...Is to Work Harder...

Many GPs are hugely conscientious – their total motivation is to do the very best they can for each of their patients.

So when a GP sees that there are still some patients in the waiting area even though the surgery is closed, rather than turning them away and saying 'please come back tomorrow' – so invoking the *repeat visit* reinforcing loop – the conscientious GP will stay in the surgery: the existence of a *backlog* or a *queue* puts *pressure on the GP to work longer hours*...

...which has the effect of increasing the *practice capacity*, at least in the short term.

The closed loop from *number of patients successfully managed per day*, through *referrals, backlogs, queues, delays and hassle* to *pressure to work longer hours* and *GP practice total capacity and capability* and back to *number of patients successfully managed per day* has a single inverse link, and so is a balancing loop – a balancing loop that acts to bring the *practice capacity* up to a level that hopefully matches the *number of cases presented per day*, so reducing the *backlog* to zero.

...but at a Price...

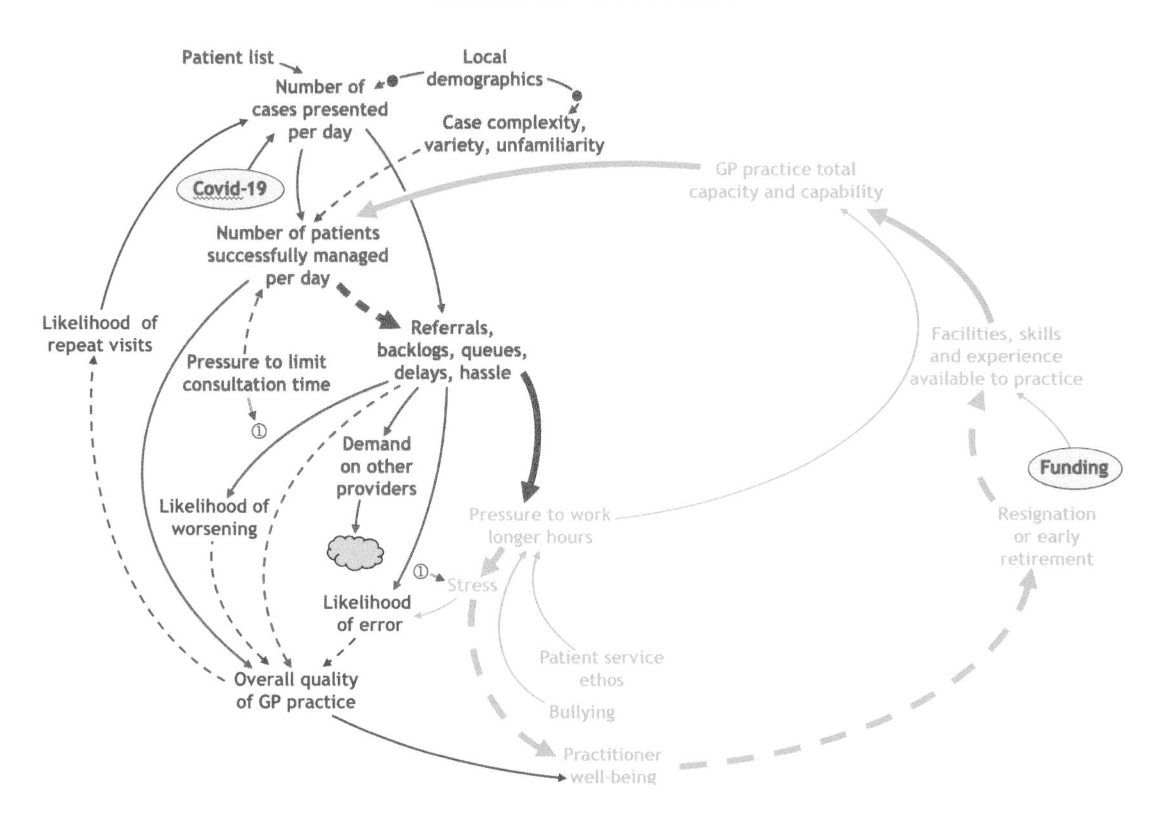

For narrative, see following page

...Which Can Be Very Damaging

Working harder can be effective in short bursts, but if anyone's working hours are prolonged day after day after day, that increases *stress*, which, in turn, can increase the *likelihood of error* as well as damaging personal *well-being. Stress* can also be a result of *pressure on the GP to limit the consultation time*, for the GP is wrestling with the tension between the 'ten minute standard consultation' and the need to spend more time with a patient to ensure that the consultation is effective.

Ultimately, a GP whose *well-being* is eroded can think 'enough is enough', so triggering *resignation or early retirement*. This depletes the *skills and experience available to the practice*, increasing the personal workload of those who remain.

This is evident from the causal loop diagram: the sequence *number of patients successfully managed per day; referrals, backlogs, queues, delays, hassle; pressure to work longer hours; practitioner well-being; resignation or early retirement; facilities skills and experience available to the practice; GP practice total capacity and capability;* and back to *number of patients successfully managed per day* is a closed loop with four inverse links, and so is a reinforcing loop. A reinforcing loop that can act as a truly vicious circle.

The diagram on page 242 shows two factors that increase the *pressure to work longer hours*, one 'honourable', the other not. We all respect a '*patient service ethos*', the GP's personal value to do the best possible for the patient. But in some environments, some people – especially those more junior – might perhaps be put under *duress* to *work longer hours*.

An Alternative Solution...

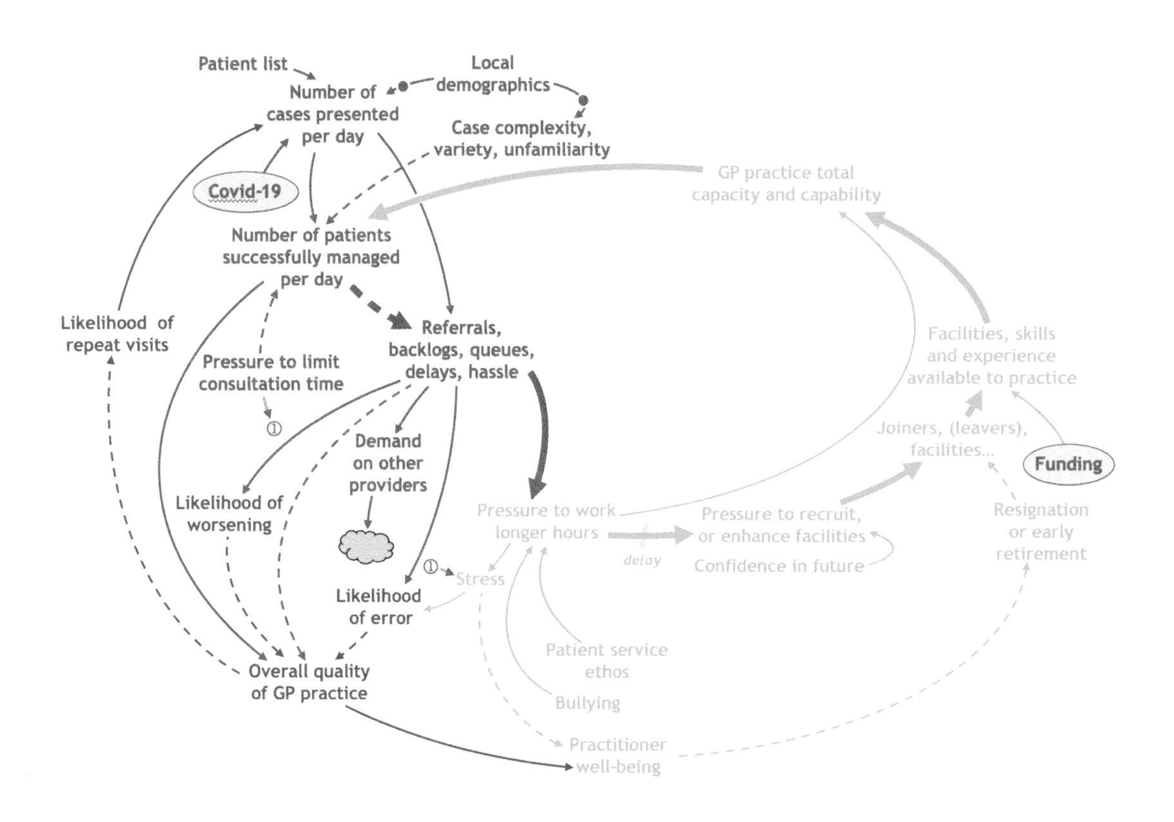

For narrative, see following page

...Is to Increase Capacity

As already noted, for the current staff to *work harder* can be effective in short bursts, but is not a long-term solution.

An alternative, and more sustainable, solution is shown in the diagram on page 244: the *pressure to work longer hours* can drive a *pressure to recruit more staff*, resulting in the appointment of *new joiners*, so increasing the *practice's skills and experience* and hence *capacity*. This completes a balancing loop that brings the *practice capacity* into line with the *number of cases presented per day*.

To *recruit more staff*, however, requires *confidence in the future*, and there can often be a *delay* in making the decision, and in finding and appointing suitable staff.

Also, *patient service* might be improved, and the *number of patients successfully managed per day* increased, by enhancing the practice's *facilities* – for example, to enable certain types of X-rays to be taken within the practice, rather than referring the patient to the local hospital.

Some Practical Considerations Relating to Recruitment…

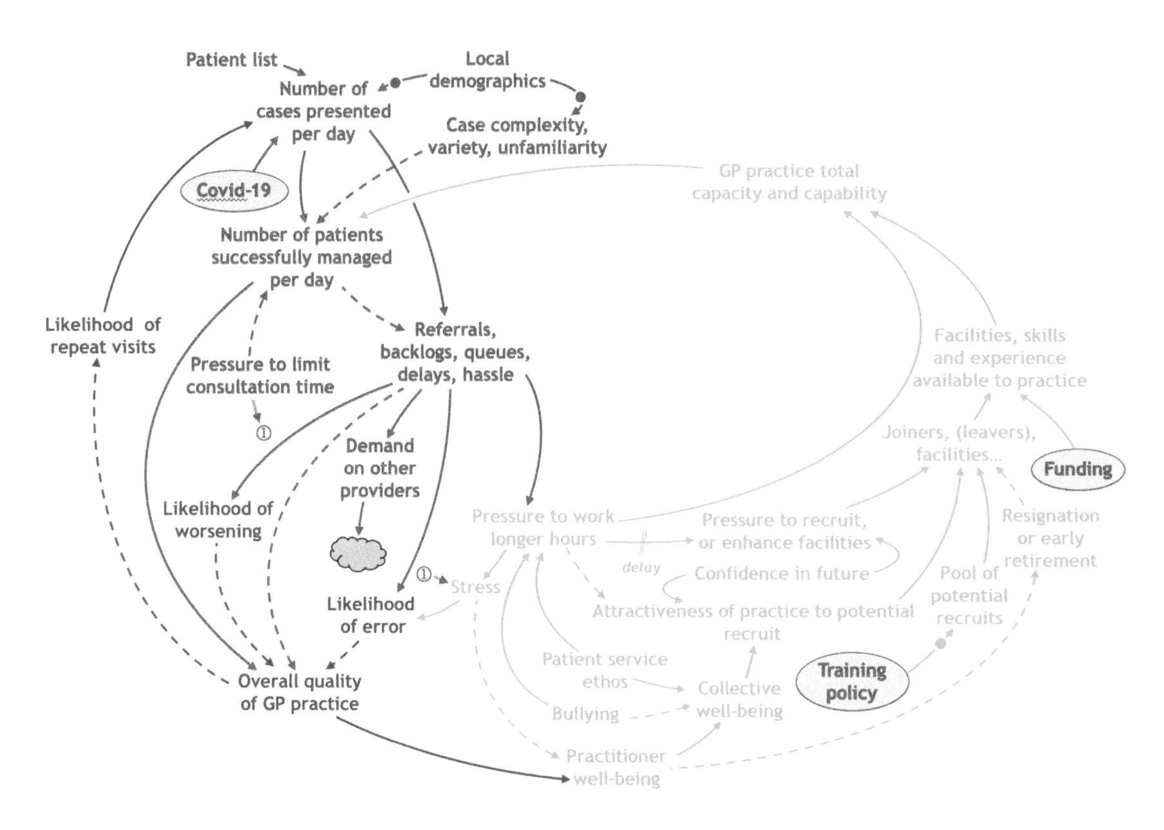

For narrative, see following page

...and Making the Practice an Attractive Place to Join

The new features shown in the diagram on page 246 identify some practical aspects of recruitment.

First, there must be a *pool of potential recruits* – people looking for a new position, or newly qualified GPs seeking their first job – which itself is influenced by government and professional *policies regarding training*, and the number of people entering the profession.

Given a *pool of recruits*, then the *practice must be attractive to a potential recruit*. As the diagram suggests, a practice that shows *confidence in the future*, and in which there is a sense of *collective well-being*, is likely to be much more *attractive* than one where all the incumbents are exhausted due to *working longer hours*, and *collective well-being* is at a low ebb.

Or Perhaps a Merger Is the Best Approach

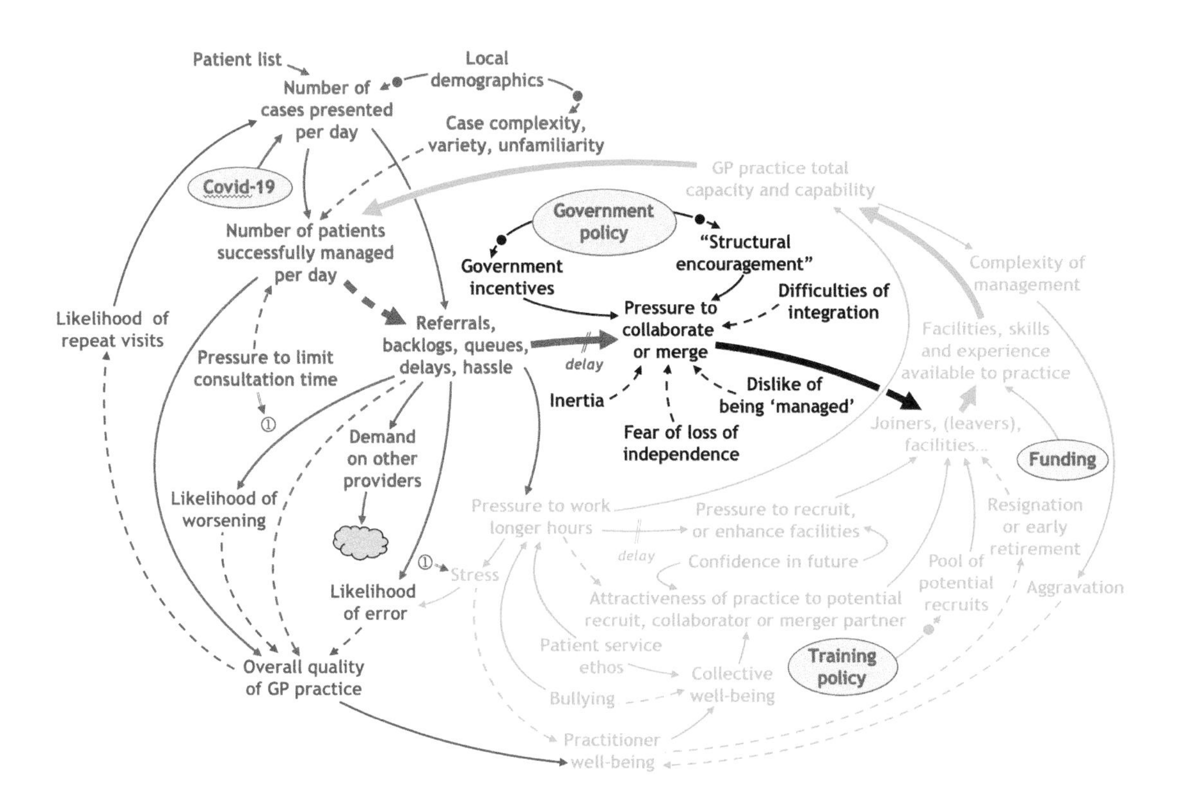

For narrative, see following page

...but Successful Mergers Aren't Easy

Recruitment brings new *joiners* into a practice gradually, one or perhaps two at a time. A more rapid way of increasing *capacity and* (especially) *capability* is for one or more practices to *collaborate*, or, more formally, *to merge*. Indeed, in England, the government has for several years offered incentives for smaller practices to merge, and the various *structures* in England that oversee health care (such as the central body known as National Health England, and the regional 'Integrated Care Systems') actively encourage smaller practices to combine to form larger delivery units.

Combining small units into larger ones can, in principle, offer benefits such as sharing resources, so increasing the *overall capacity and capability*, and having a more stable financial base, which might be able to fund an investment (such as the purchase of an X-ray machine, and the employment of suitably skilled staff to operate it) that a single smaller practice could not afford.

Achieving these benefits, however, does not 'happen by itself', but requires active management, takes time, and can be disruptive. The anticipation of the *difficulties of integration* can therefore act as a powerful disincentive, slowing down and perhaps jeopardising, discussions between potential merger partners.

Personal preferences and attitudes are important too. Someone used to running a small practice, perhaps as the sole partner or with one or two others, might be very reluctant to contemplate being one partner among, say, 20, *fearing a potential loss of personal independence*, and, more generally, *not relishing the prospect of having to report, managerially, to someone else* after having been 'the boss' for so many years past.

Furthermore, larger business units are inherently more *complex to manage* than smaller ones, with chains of accountability, perhaps stricter rules and procedures, and (almost certainly!) more forms to fill in. For those who prefer more 'flexible' ways of working, in which the annual appraisal is a drink in the local pub, working as part of a larger unit offers nothing but *aggravation*...

The Bigger Picture

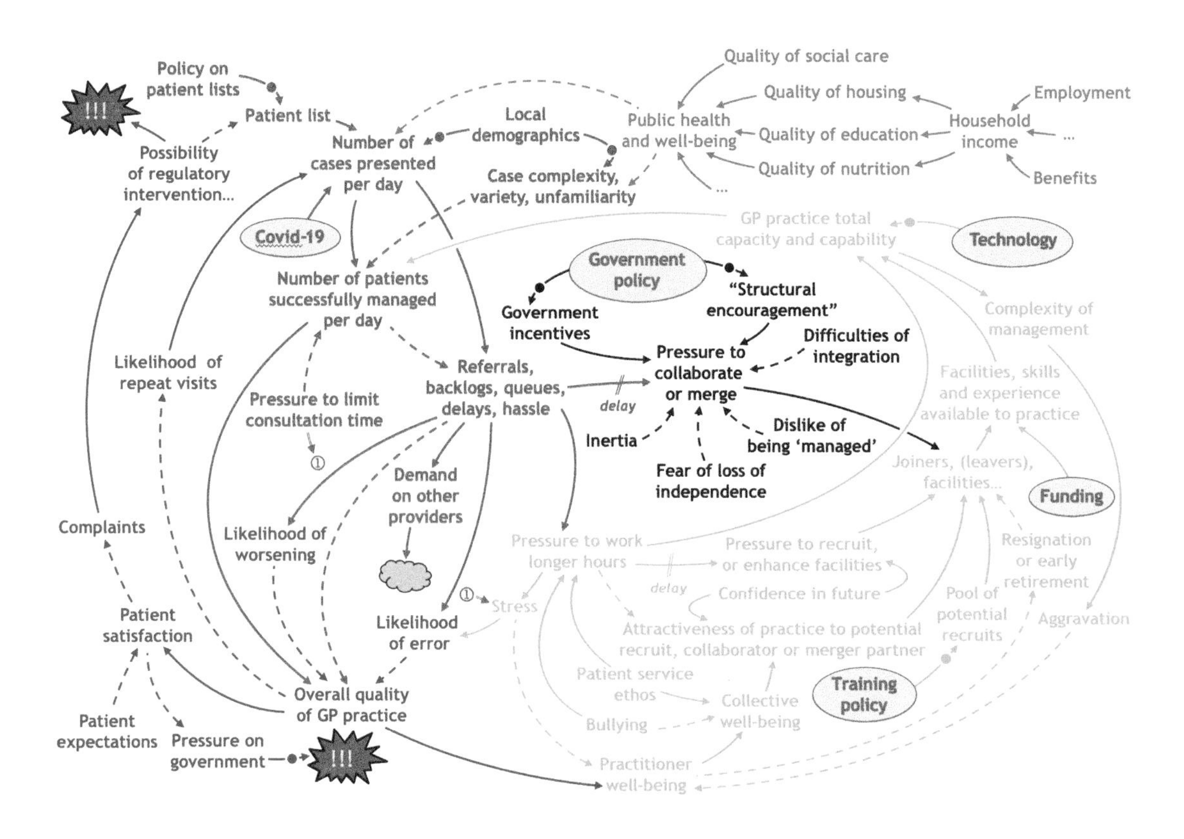

For narrative, see following page

GP Practices in Context

The diagram shown on page 250 puts the GP practice into a wider context.

On the left are some aspects relating to the patient, the key feature of which is the *patient's satisfaction*, this being the individual patient's assessment of their treatment, as determined by comparing their perception of the *quality of the practice* against their *expectation*: if the *quality* exceeds the *expectation*, the patient is likely to be (at least reasonably) satisfied; but if the *quality* does not meet the *expectation*, then the patient could well be displeased.

The lower the *quality of the practice* – resulting, perhaps, from *backlogs and queues* – the lower the *patient satisfaction*, perhaps to the extent that the patient raises a *complaint*. This, in turn, could cause the appropriate *regulatory body to intervene*, perhaps resulting in some form of investigation or enquiry, which might exonerate the individual GP and the practice as a whole. But it might not – leading to perhaps a fine, or a disciplinary tribunal, or – in the case of a GP practice – the transfer of the *patient list*.

And if many patients, across the country, are dissatisfied, then this can put *pressure on the government*. Which, during the Covid-19 crisis in the UK, the government understood very well. Throughout the crisis, the government's over-riding priority was 'to protect the National Health Service'. Yes, that is good for the general population. But it is also absolutely necessary for the government's survival – were the UK National Health Service to be overwhelmed, the government would surely fall.

Within the practice itself, *technology* can have a substantial impact on the way in which a GP's services are delivered, as indeed has been experienced in the Covid-19 crisis, during which there has been an increased use of remote, rather than face-to-face, consultations.

Across the top, in magenta towards the right, are just some of the factors that influence the *number of cases presented per day* and the *case complexity and variety*. *Housing, education and nutrition* (each of which, directly or indirectly, is determined by the *household's income*), as well as *social care*, all affect *public health and well-being*, and a *household's income* is determined by the local *employment opportunities*, the availability and level of *benefits*, the *educational levels* of income earners, the extent to which members of the household have to act as *carers* for others...

Yes, it's all very complicated. And it all operates on vastly different scales – from an individual GP partner feeling rather grumpy about the prospect of *being 'managed'* as the result of a proposed merger, to government policy on incentives to new industries to provide *employment* in an area that no longer builds ships.

The Referral Chain

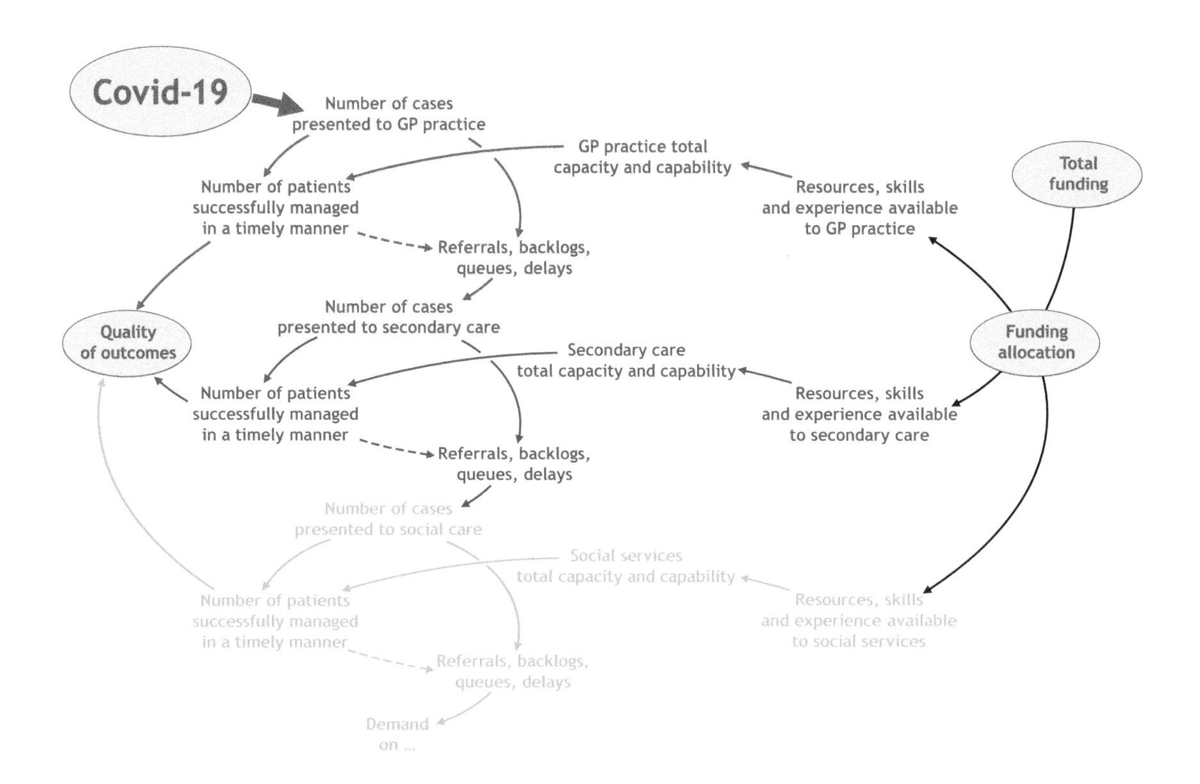

For narrative, see following page

One Level's Referral Is the Next Level's Demand

On page 237, the discussion of the diagram on page 236 states that if a GP practice does not have the *capability* to provide a full and successful treatment for a given *case as presented*, then a *referral* is made elsewhere – for example, the GP might write a prescription for medication which can be taken to a pharmacist, or arrange a diagnostic test at a laboratory, or recommend the patient for admission to a hospital. In all the subsequent diagrams, the resulting *demand on other suppliers* was represented as a 'cloud'.

In fact, the prescription, the request for a test, or the hospital admission, are each the equivalent of *a case presented on a given day to a GP practice*, as experienced by the pharmacy, laboratory or hospital, each of which has its own appropriate *capacity and capability*, and its own pattern of demand, to which the *referral from the GP practice* adds.

The diagram shown on page 252 therefore represents a (highly simplified) view of the 'referral chain', for the particular example of a GP practice that refers a patient to a hospital, so placing a demand on the hospital's services, requiring for example, a suitable bed and appropriate medical attention. Then, following treatment in the hospital, the patient is discharged, and requires some care at home, as provided by the locality's social services.

The *capacity and capability of the GP practice* can therefore influence the 'downstream' requirement for *social care*. So, for example, it is quite possible that a *backlog* at the GP practice causes a delay in treating a particular patient, so causing the condition to worsen, so that by the time the patient is seen, the condition can no longer be fully treated by the GP, and therefore requires admission to a hospital. And because the patient is elderly, full recovery after leaving the hospital requires some social care at home. Social care that would not have been necessary had the patient been seen in a timely manner at the outset. And when the overall system is subject to a shock of the scale of Covid-19, the consequent disruption can be immense: patients wait longer to see their GPs, hospitals cancel 'non-urgent' operations, social care is overwhelmed…

The diagram shown on page 252 is a simplification, but I trust an insightful one. And a general one too, for it applies to all 'referral chains', for example within a hospital (admissions to the Accident and Emergency Department place a demand for beds in wards, and also perhaps on the X-ray department and the operating theatre…) as well as other contexts (one instance being a customer service call centre in which the first point-of-contact is unable to deal with the customer enquiry, and so gives the enquirer another number, which, when called, doesn't resolve it either…).

Yes, it's all very complex. As are the fundamental decisions concerning best to allocate funds across the whole 'system'. And that's why systems thinking can really help. The complexity is there, and won't 'just disappear'. So far better to try to capture, and tame, the complexity by compiling a causal loop diagram, for that provides a well-evidenced and robust platform for discussion, re-design and creativity.

How These Causal Loop Diagrams Can Be Used

The causal loop diagrams shown on pages 248, 250 and 252 tell a disturbing – but none the less true – story, a complex story full of problems. But what they don't do is give answers, solutions as to how those problems can be satisfactorily addressed. And of course it's the solutions that we want – and so some people feel somewhat disappointed that the causal loop diagrams don't deliver them. Any disappointment, however, is misplaced: in the first instance, the key purpose of a causal loop diagram is to capture, vividly and lucidly, the essence of the-way-things-are-right-now, so that we can all be 'on the same page'. And if the-way-things-are-right-now is problematic, those problems will be seen, starkly and powerfully. This then provides a sound basis for discovering solutions, for any solution *must be some form of intervention in the system-as-it-is-now to make something DIFFERENT happen.*

The 'as-is' causal loop diagram is therefore a trigger for asking a range of perceptive questions, all of the generic form 'How might [this] be different?', as a result of which ideas will be generated, and solutions discovered. Accordingly, with reference to the diagram on page 248, a GP practice might ask questions such as:

- What can we do differently to increase our capability and capacity?'
- What can we do differently to make more, and better, use of technology?'
- What can we do differently to make our practice even more attractive to potential recruits?'
- What might we do differently to improve communication with, for example, local hospitals, pharmacists, and others to whom we refer our patients?'

With reference to the diagram on page 252, a healthcare agency that has authority over GP practices, hospitals and social care across a region might ask:

- How can our funding allocation be different to ensure smoother flows across all levels, with fewer backlogs, queues and delays?'

And, back to the diagram on page 250, a government might ask:

- What needs to be different about our current policies so as to improve overall public health and well-being so that the demand on healthcare and related services is fundamentally reduced?'

If you want to 'make the world a better place', something about world-as-it-is-now has to be different. Systems thinking, and an insightful causal loop diagram, will help you ask the right questions, so enabling good answers to be discovered.

Chapter 17

The Climate Crisis

DOI: 10.4324/9781003304050-19

The Climate *Is* a Crisis

The man-made climate crisis is the most significant threat, not only to mankind, but to the entire planet. The 'global climate system' is the most complex system that exists, and so cannot be 'controlled' or 'managed' by any one nation or corporation. So the only way to 'manage' it is by acting collectively, by showing true international co-operation, teamwork and trust – as, for example, discussed on pages 160 to 163. And each of us, as individuals, and in the roles we play within any organisations of which we are a part, can influence how that system behaves, for we are all 'entities' within it.

To influence the climate system beneficially, however, we need to have an understanding of how the climate system behaves, and so this chapter present a series of causal loop diagrams that does just that.

And the outcome might be a surprise.

Yes, reducing carbon-based emissions is indeed a 'good thing' to do.

But as the following pages will show, reducing emissions alone will not solve the fundamental problem. To do that, something else is required – 'something else' that emerges directly from the causal loop diagrams.

And showing that the world's most complex system can be succinctly and informatively described in a causal loop diagram on a single page (as will be seen on page 288) must be the most convincing demonstration of the power of systems thinking, and the use of causal loop diagrams, in taming the complexity of real-world problems.

'But It Is Increasingly Clearer That Reducing Emissions Is Not Enough – We Must Also Actively Remove Greenhouse Gases from the Atmosphere'

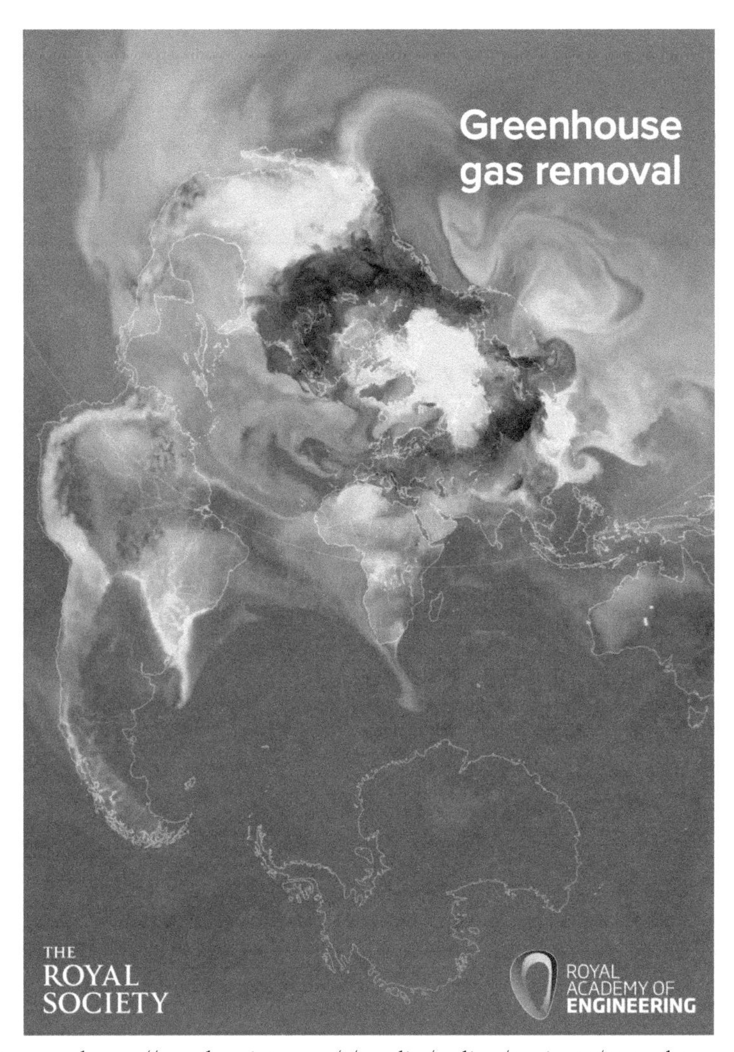

Source: https://royalsociety.org/-/media/policy/projects/greenhouse-gas-removal/royal-society-greenhouse-gas-removal-report-2018.pdf, page 7

Handprints in the Cueva de las Manos, Argentina, c. 10,000 Years Old

Source: https://upload.wikimedia.org/wikipedia/commons/f/f4/SantaCruz-CuevaManos-P2210651b.jpg

The Story of Man Over 10,000 Years...

As we saw on pages 56–58, the reinforcing loop of *births* seeks to grow exponentially, as determined by the *birth rate*. But at the same time, the *global population* is being depleted by *deaths*, as determined by the *death rate*.

The dynamic behaviour of this system – a reinforcing loop seeking to grow exponentially, constrained by a balancing loop – can be very complex, and depends on the instantaneous behaviours of the *birth rate* and the *death rate*. If, at any time, the *birth rate* exceeds the *death rate*, the *population* grows; if the *death rate* exceeds the *birth rate*, the *population* declines; if the *birth rate* equals the *death rate*, the *population* is stable.

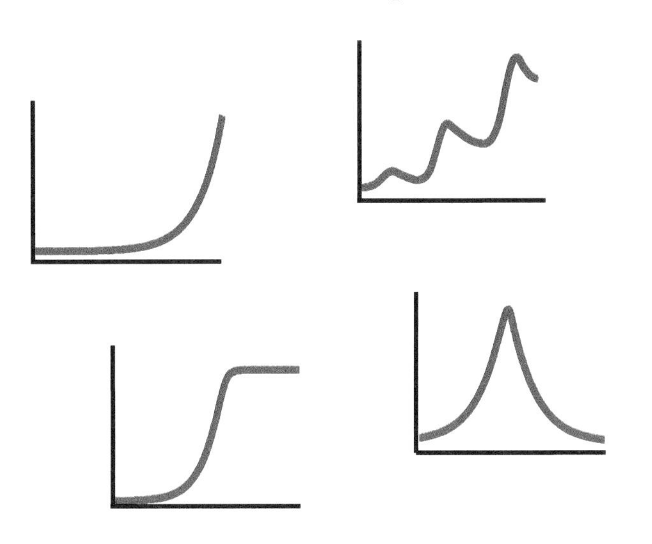

In the graphs to the left, the vertical axis represents *population*, and the horizontal axis, time. Each shows a possible behaviour over time of the *population*, depending on the instantaneous values of the *birth rate* and the *death rate*.

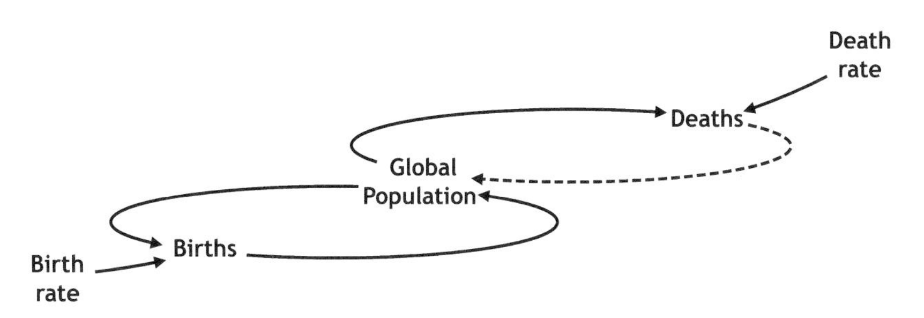

Human Activity…

Driven by the needs for *survival and the desire for wealth*, a *population* will engage in *human activity*, such as farming, building, manufacturing and trading. All these activities *consume resources* such as land and water, minerals and gases, coal and oil, resulting in *pollution* (such as carbon dioxide from fossil fuels, and nitrogen oxides and particulates from diesel engines), *waste* (such as plastics, CFCs and industrial waste) and *by-products* (such as methane from farm animals).

Many resources have only a finite *total capacity*, implying that the *available resource capacity* becomes progressively depleted as a result of *consumption*, eventually leading to *competition for resources. Pollution*, *waste* and *by-products* can also reduce the *available resource capacity*, for example, as a result of contamination.

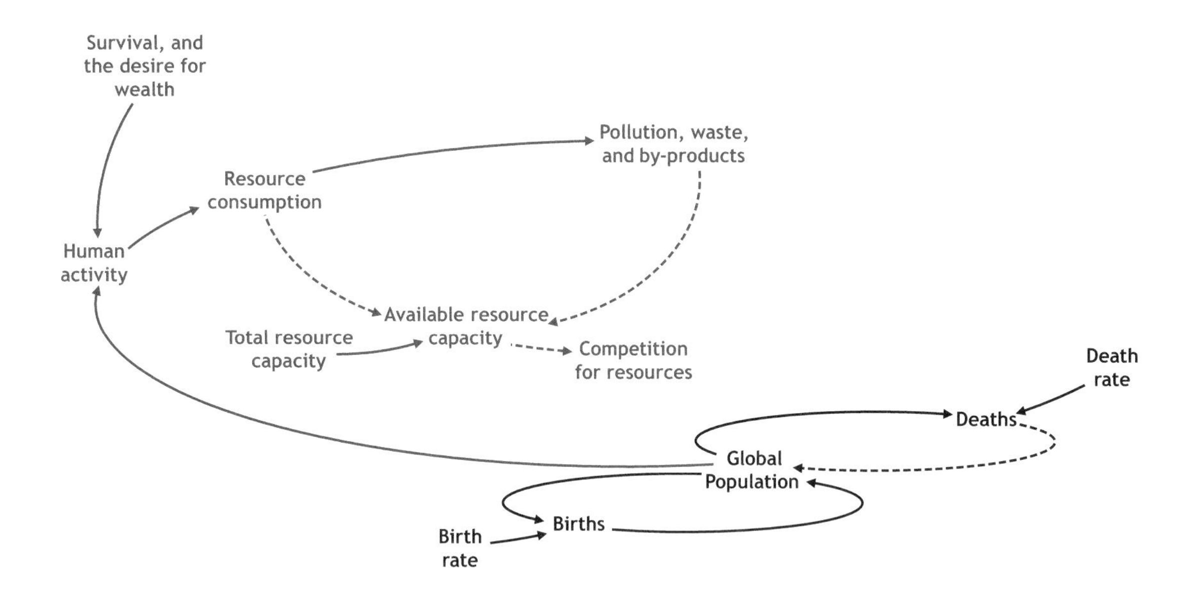

...Leads to War, Famine and Disease

Pollution, waste and by-products create the conditions in which *disease* flourishes, and *competition for resources* can result in *war* and *famine*. *War*, *famine* and *disease* all increase the *death rate* beyond that attributable to natural processes such as ageing.

The closed loop from *population* through *human activity*, *resource consumption*, *pollution* and *disease* round to *death rate*, *deaths* and back to *population* has only one inverse link. One is an odd number, and so this loop is a balancing loop, which acts to arrest the growth of the *births* reinforcing loop. This is just one of six balancing loops, all of which act simultaneously to limit the *births* reinforcing loop.

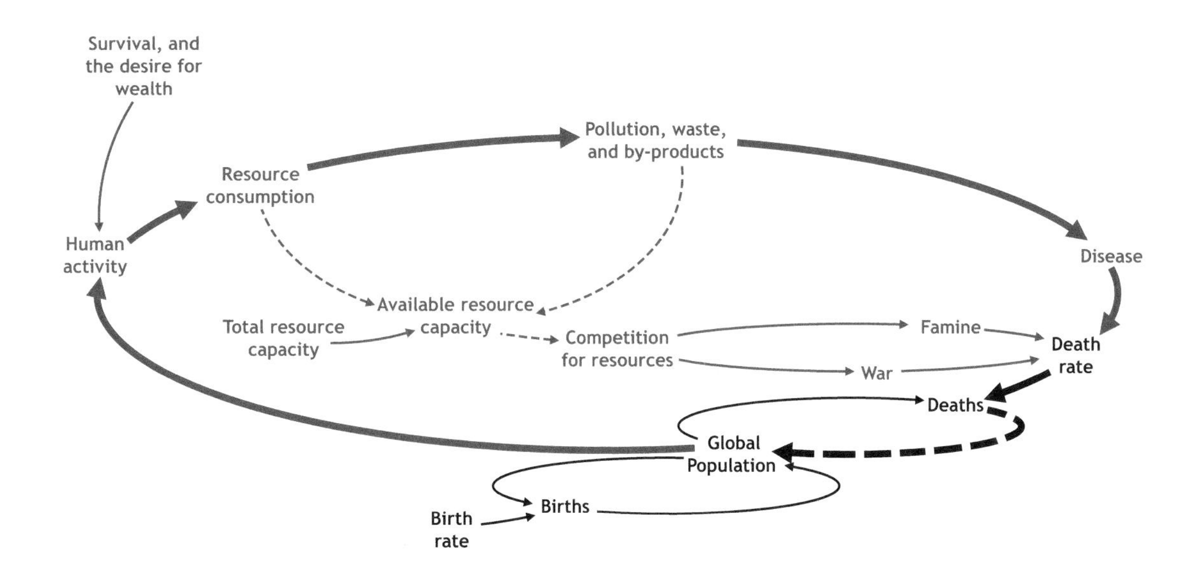

The Four Horsemen

The diagram on page 261 shows a single reinforcing loop seeking to grow exponentially, being limited by six simultaneous balancing loops, all resulting in *deaths* – either from natural causes, or resulting from *disease*, *famine* and *war*.

The Four Horsemen of the Apocalypse – pestilence, war, famine, death – are real, and for many hundreds, if not thousands, of years, held the global human population in check, with very slow growth.

https://upload.wikimedia.org/wikipedia/commons/2/29/Durer_Revelation_Four_Riders.jpg

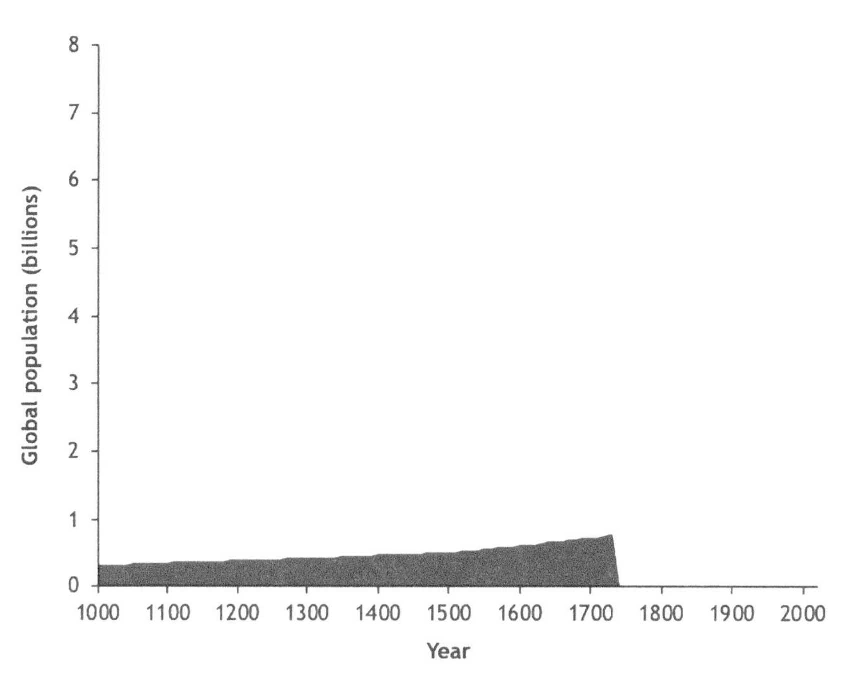

Source: The United Nations Population Division, Department of Economic and Social Affairs

The Benefits of Public Health

As economies became more developed, and our understanding of *disease* became more scientific, some *human activity* was devoted to *public health*, especially the provision of clean water, and the treatment of sewage, in cities. This has two, simultaneous effects: first, improved *public health* increases the *birth rate*, due to both a reduction in infant mortality and the enhanced health of women of child-bearing age, and second, improved *public health* decreases the incidence of *disease* (for example, water-borne diseases such as cholera), so reducing the *death rate*.

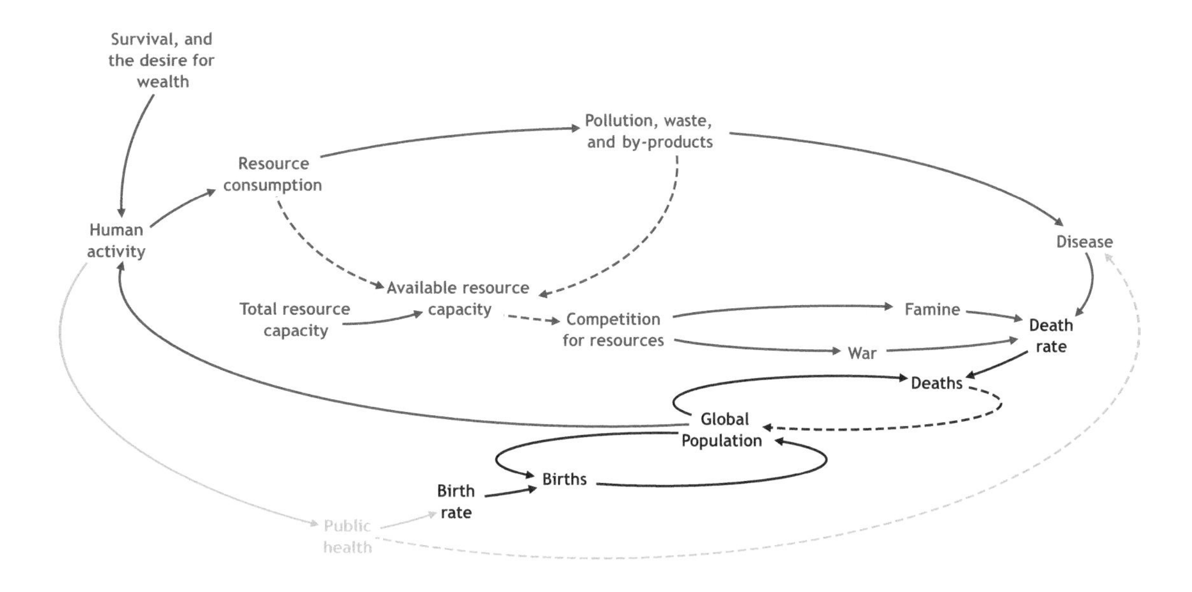

The Population Explosion

As we have already seen, the rate of growth of the *population* is driven by the difference between the *birth rate* and the *death rate*, so suppose that, before the focus on *public health*, the *birth rate* is ten live births per 1,000 people, and that the *death rate* is nine deaths per 1,000 people. The net *birth rate* is therefore 10−9=1 person per 1,000 people, and so the *population* will grow at this rate.

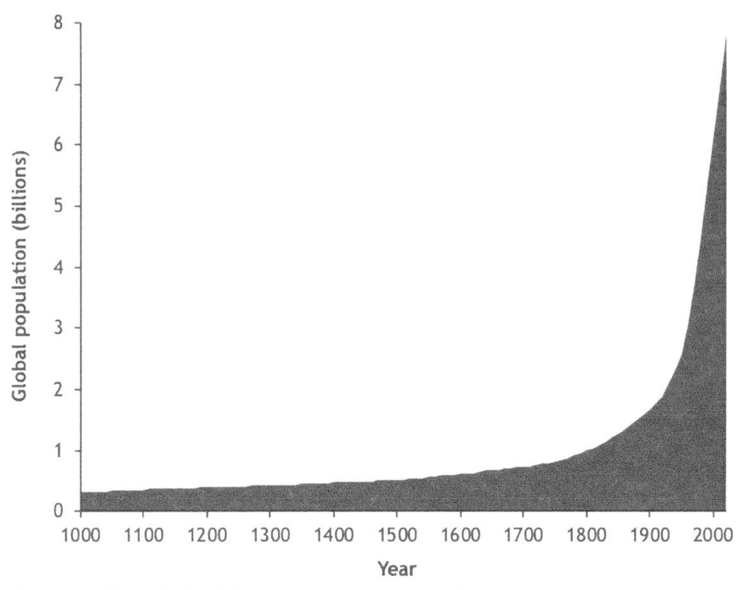

Source: The United Nations Population Division, Department of Economic and Social Affairs

Suppose further that a *public health* programme has the result of reducing the *death rate*, and increasing the *birth rate*, each by about 10%. The *death rate* therefore decreases from 9 to 8 deaths per 1,000 people, and the *birth rate* increases from 10 to 11 live births per 1,000 people. The net *birth rate* therefore changes from 10−9=1 person per 1,000 people to 11−8=3 per 1,000 people. A 10% change to each of the *birth rate* and the *death rate* causes a 300% increase in the growth rate of the *population*.

As can be seen from the chart to the left, the *population* explodes.

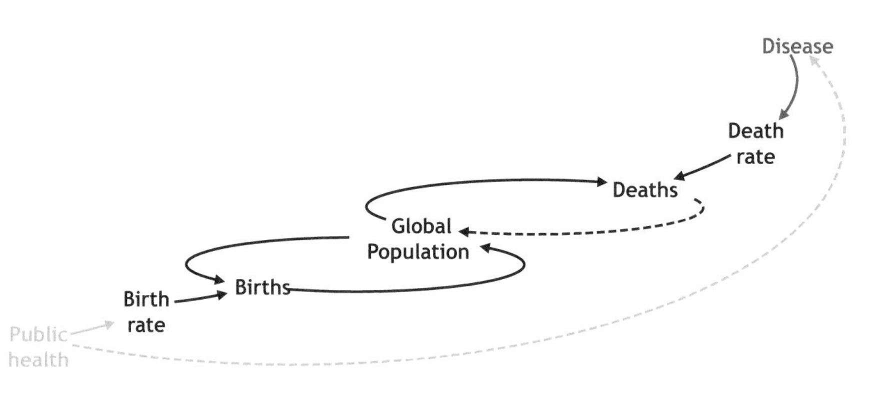

The Story so Far...

The most effective current way of limiting the *birth rate* is by devoting some of the *human activity* to the *education of women*, but this can take a long time, and has significant cultural implications.

From a systems perspective, this introduces another balancing loop – *population* to *human activity* to *education of women* to *birth rate* to *births* and back to *population*. This limits the growth of the *births* reinforcing loop, but in a far more humane way than as a result of *disease*, *famine* and *war*.

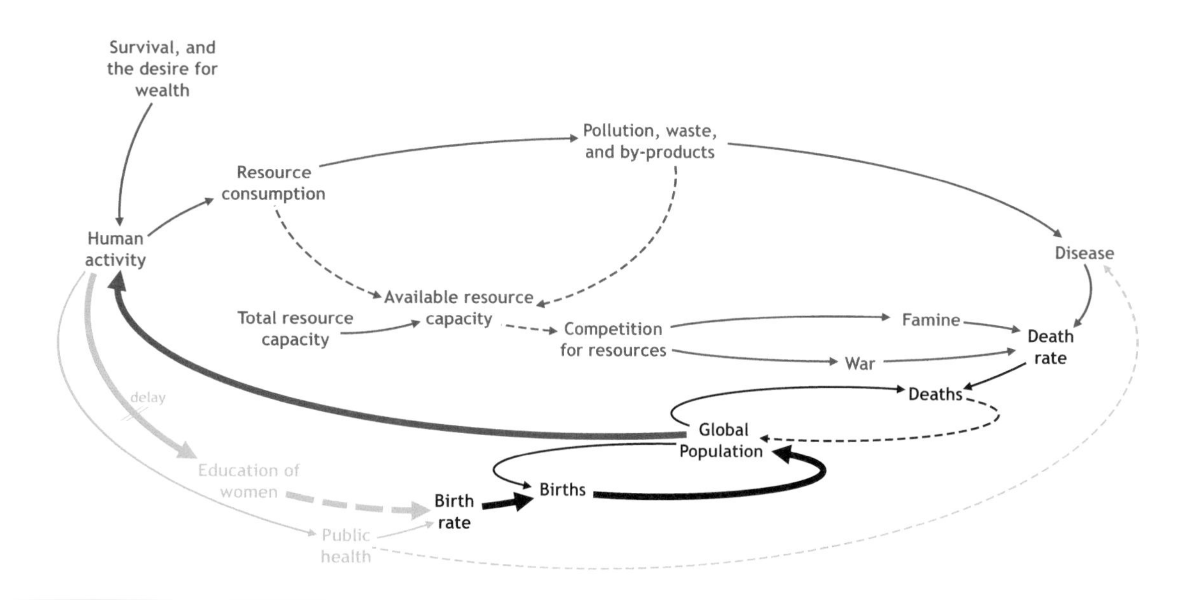

James Lovelock, 1919 – 2022

Source: https://upload.wikimedia.org/wikipedia/commons/4/44/James_Lovelock_in_2005.jpg

Why Isn't the Earth Getting Hotter?

If you heat a block of metal, it's gets hotter. But not indefinitely – sooner or later, the metal's temperature stabilises. This happens because the metal's temperature is the net result of two opposing effects: the rate at which energy is absorbed by the metal from the external heat source, and the rate at which energy is lost from the metal as a result of its own radiation. If more energy is absorbed than is lost, the metal gets hotter, as happens when the metal is relatively cool. But as the temperature of the metal increases, the rate of energy loss also increases, until the rate at which energy is lost equals the rate at which energy is absorbed, at which point the temperature of the metal stabilises.

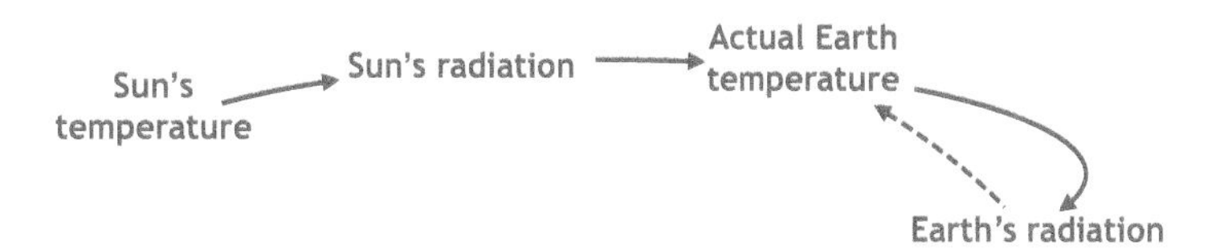

The Earth is rather like a block of metal, with the heat source as the sun. The *sun's radiation,* attributable to the *sun's temperature*, heats the Earth, so increasing the *actual Earth temperature*. But as the Earth warms, the intensity of the *Earth's radiation* increases too, progressively slowing down the rate at which the *actual Earth temperature* rises. This forms a balancing loop, such that when the *actual Earth temperature* has risen to a value at which the rate of heat loss attributable to the *Earth's radiation* equals the rate of heat input from the *sun's radiation*, the *actual Earth temperature* stabilises.

According to this balancing loop, the stabilised *actual Earth temperature* should track the *sun's temperature*: if the sun becomes warmer, the Earth should become warmer too; if the sun becomes cooler, the Earth should become cooler.

Over the last several billion years, the sun has become increasingly hotter – that's what happens to all stars as they 'age'.

According to this balancing loop, the *actual Earth temperature* should have been rising steadily too.

But has it?

There Must Be Another Balancing Loop

No. It hasn't. Geologic evidence shows that the *actual Earth temperature* has been more-or-less constant at about 14°C for billions of years, whilst, over that time, the intensity of the *sun's radiation* has been increasing. This cannot be explained by the balancing loop shown on page 267 alone. Rather, it suggests the presence of *another* balancing loop acting to stabilise on the *'natural' Earth temperature*, 14°C.

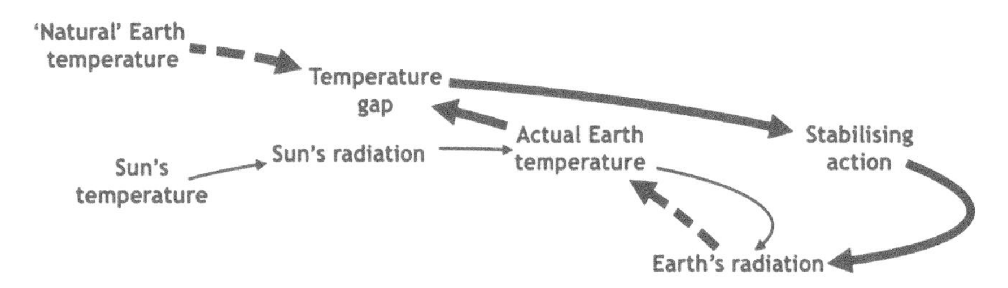

The action of this second balancing loop is best understood by considering what happens as the intensity of the *sun's radiation* increases, causing the *actual Earth temperature* to rise above the *'natural' Earth temperature*. This opens a *temperature gap* which triggers some type of *stabilising action* to increase the intensity of the *Earth's radiation*, so reducing the *actual Earth temperature*. This then closes the *temperature gap*, and brings the *actual Earth temperature* into line with the *'natural' Earth temperature*.

This form of temperature control is similar to that which we experience in our own bodies: when we are too hot – when our actual temperature exceeds our 'natural' temperature of 36.9°C – we invoke the stabilising action of sweating, which acts to reduce our actual temperature back to the 'natural' value.

But what is the Earth's stabilising action?

The answer lies in James Lovelock's Gaia theory...

Planetary Atmospheres

In the 1960s, James Lovelock was studying the atmospheres and surface temperatures of Venus, the Earth and Mars. A significant feature of these planetary atmospheres concerns their composition: the Earth's atmosphere is about 20% oxygen and 80% nitrogen, very different from the atmospheres of Venus and Mars which are almost entirely carbon dioxide. Lovelock knew that oxygen is very chemically reactive: a strike of lightning, for example, can spark a forest fire in which atmospheric oxygen reacts with materials in the wood to form carbon dioxide. Why is there so much unreacted oxygen in the Earth's atmosphere? According to the laws of science, the oxygen should have reacted – and all trees should have been consumed by fire – long ago. Lovelock then calculated what the Earth's atmosphere would look like if all the oxygen had reacted, such that the atmosphere were at what chemists call 'chemical equilibrium'. His result is shown in the right-hand column of the chart below and to the left: if the Earth's atmosphere were at chemical equilibrium, its composition would be very similar to the atmospheres of Venus and Mars.

Lovelock also estimated what the temperature of the Earth would be if the atmosphere were at equilibrium. As shown below at the far right, that temperature would be around 250°C. But the actual Earth temperature is much, much, lower – about 14°C.

All of which posed some important questions. Why is the Earth's atmosphere not at chemical equilibrium? Why is the Earth's temperature lower than it 'should' be? And are the answers to these two questions related?

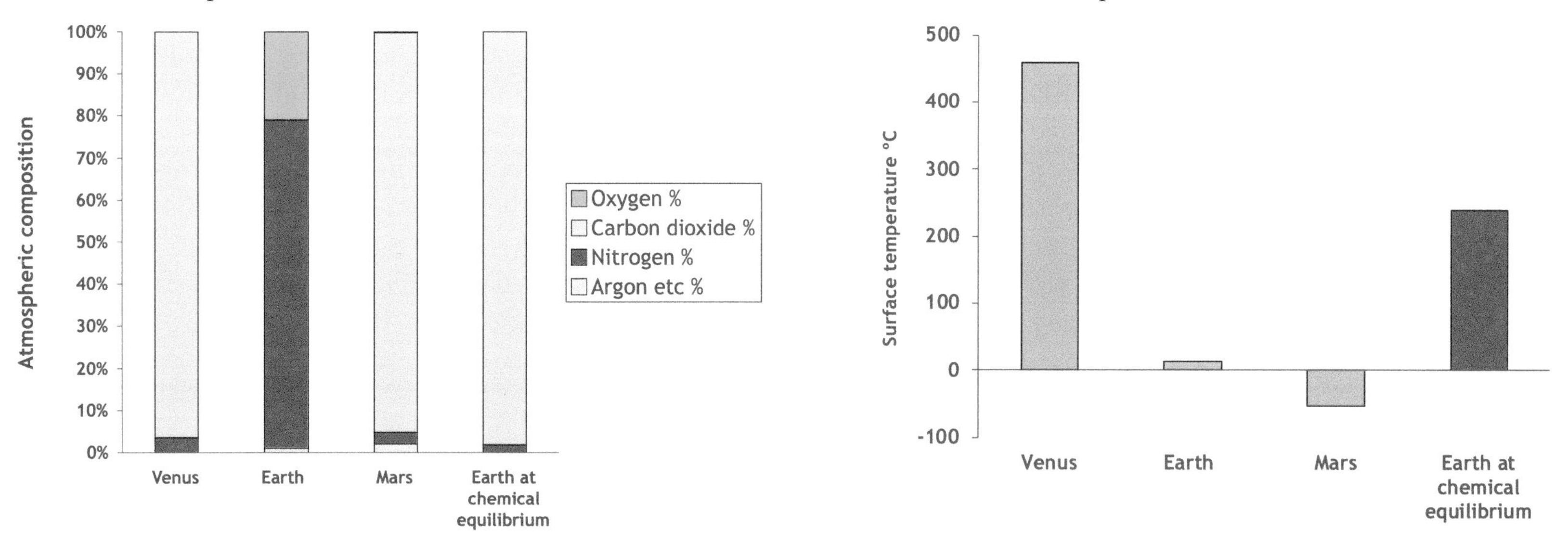

Source: The Ages of Gaia, by James Lovelock, published by Oxford University Press, 2nd edition 1995, page 9.

Gaia

Lovelock's key observation was that the Earth maintains a stable state, far from chemical equilibrium. But he'd seen this before. As you and I have. For you and I are systems far from chemical equilibrium, and systems which maintain a stable state. One example of this is how human beings maintain a stable internal temperature of 36.9°C. If our environment is hot, we sweat, or we increase the flow of blood to our skin so as to enhance heat loss; if our environment is cold, we shiver, or decrease the flow of blood to our skin so as to reduce heat loss. As a result, our internal body temperature is maintained stable at 36.9°C even if our immediate environment is warmer or cooler.

Lovelock's great insight as to why the Earth can maintain itself away from chemical equilibrium is because the Earth as a whole – its structure, its rocks, its oceans, its weather, its living beings – collectively behave as a living 'super-organism', which he named 'Gaia'. And as a 'super-organism' (more formally, as a self-organising complex system), Gaia acts to maintain the conditions necessary for survival, such as a stable Earth temperature.

He also identified the primary mechanism by which the Earth keeps its temperature stable at 14°C, despite the increasing intensity of the sun's radiation. The key factor is the quantity of carbon dioxide in the atmosphere. As a 'greenhouse gas', carbon dioxide acts as an atmospheric 'blanket' such that the greater the quantity of carbon dioxide in the atmosphere, the warmer the Earth.

For us to keep cool, we sweat; for the Earth to keep cool, carbon dioxide needs to be removed from the atmosphere.

Page 271 shows the mechanism Lovelock identified. Carbon dioxide in the air dissolves in rain to form a weak solution of carbonic acid. When the rain falls to Earth, the carbonic acid reacts with calcium silicate in rocks to form silicic acid and calcium bicarbonate. This is known as the 'weathering of rocks', and is much speeded up by microorganisms in the soil. Calcium bicarbonate is soluble, and flows into the sea, where algae known as 'coccolithophores' absorb it, using the energy of sunlight to transform it into calcium carbonate. Calcium carbonate is insoluble, and forms shells around the living algae. When the algae die, the calcium carbonate shells fall to the bottom of the sea, and over geological time, form the rocks we know as chalk and limestone.

All the chalk and limestone in the world originates from this process, which has the effect of 'pumping' carbon dioxide out of the air, and 'burying' it as rock – a life-mediated process which Lovelock called 'the living pump'. And by removing carbon dioxide from the atmosphere, the Earth can keep cool.

The Living Pump

Schematic representation of the living pump

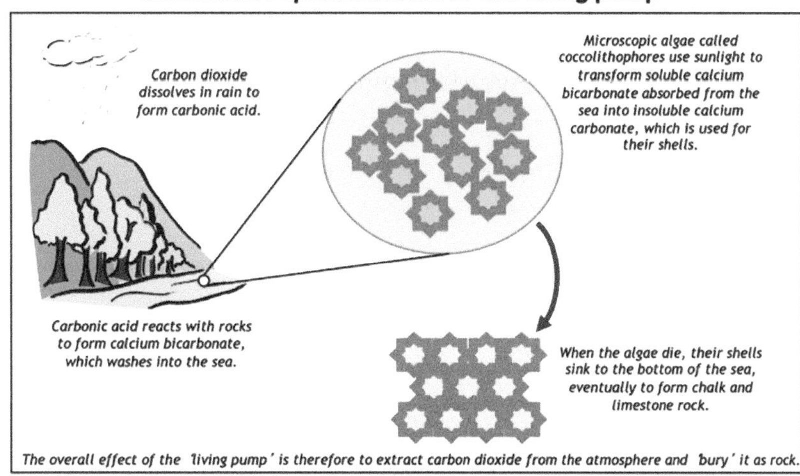

A 'bloom' of coccolithophores at the western entrance to the English Channel, between Cornwall and the Cherbourg peninsular

Landsat imagery courtesy of NASA Goddard Space Flight Center and U.S. Geological Survey
https://eoimages.gsfc.nasa.gov/images/imagerecords/146000/146897/englishchannel_tmo_2020176_lrg.jpg

Chalk cliffs at Møns klint, Denmark, formed from the shells of countless billions of coccolithophores

From image courtesy of Mads Sabroe
https://commons.wikimedia.org/wiki/File:Møns_klint_-_panoramio.jpg

Scanning electron micrographs of three coccolithophores
Images reproduced courtesy of William Balch, Bigelow Laboratory for Ocean Sciences
https://www.bigelow.org/files/books/coccolithophorebook.pdf

Controlling the Earth's Temperature

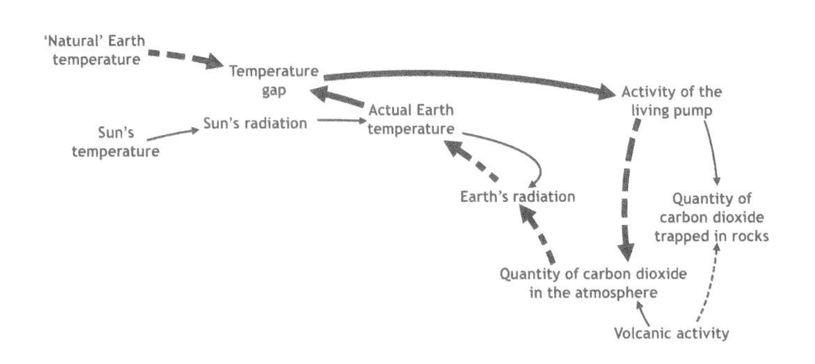

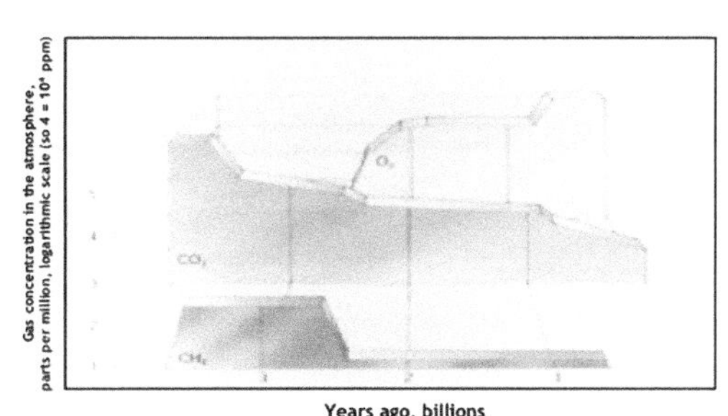

Source: ***Gaia: The practical science of planetary medicine*, James Lovelock, Gaia Books Limited (1991), page 113**

We can now answer the question posed on page 268 – what is the *stabilising action* that has maintained the *actual Earth temperature* more-or-less constant at the *'natural' Earth temperature*, 14°C, for billions of years? According to Gaia theory, the stabilising action is the *activity of the living pump*.

When the *actual Earth temperature* rises above the *'natural' Earth temperature*, as caused by an increase in the intensity of the *sun's radiation*, the *temperature gap* stimulates the *activity of the living pump*, so reducing the *quantity of carbon dioxide in the atmosphere* and simultaneously increasing the *quantity of carbon dioxide trapped in rocks*. Reducing the *quantity of carbon dioxide in the atmosphere* increases the intensity of the *Earth's radiation*, and so reduces the *actual Earth temperature* to close the *temperature gap*.

This loop has three inverse links, and so is, as expected, a balancing loop, stabilising on the *'natural' Earth temperature*.

For completeness, this causal loop diagram also shows the effect of *volcanic activity*, which releases carbon dioxide trapped in rocks back into the atmosphere. Although the living pump and volcanoes work against one another, until very recently, the living pump worked fast enough to counteract the effect of volcanoes.

As shown in the chart on the right, over billions of years, as the sun has been getting hotter, the living pump has steadily reduced the quantity of carbon dioxide in the atmosphere, so keeping the Earth's temperature at about 14°C.

Man *v.* Gaia

Who Will Win?

Direct Air Capture

The Climeworks 'Orca' installation for the direct capture of atmospheric CO_2, as installed at the Hellisdeiði power plant in Iceland

Image reproduced courtesy of Climeworks, https://climeworks.com/.

For a Long Time…

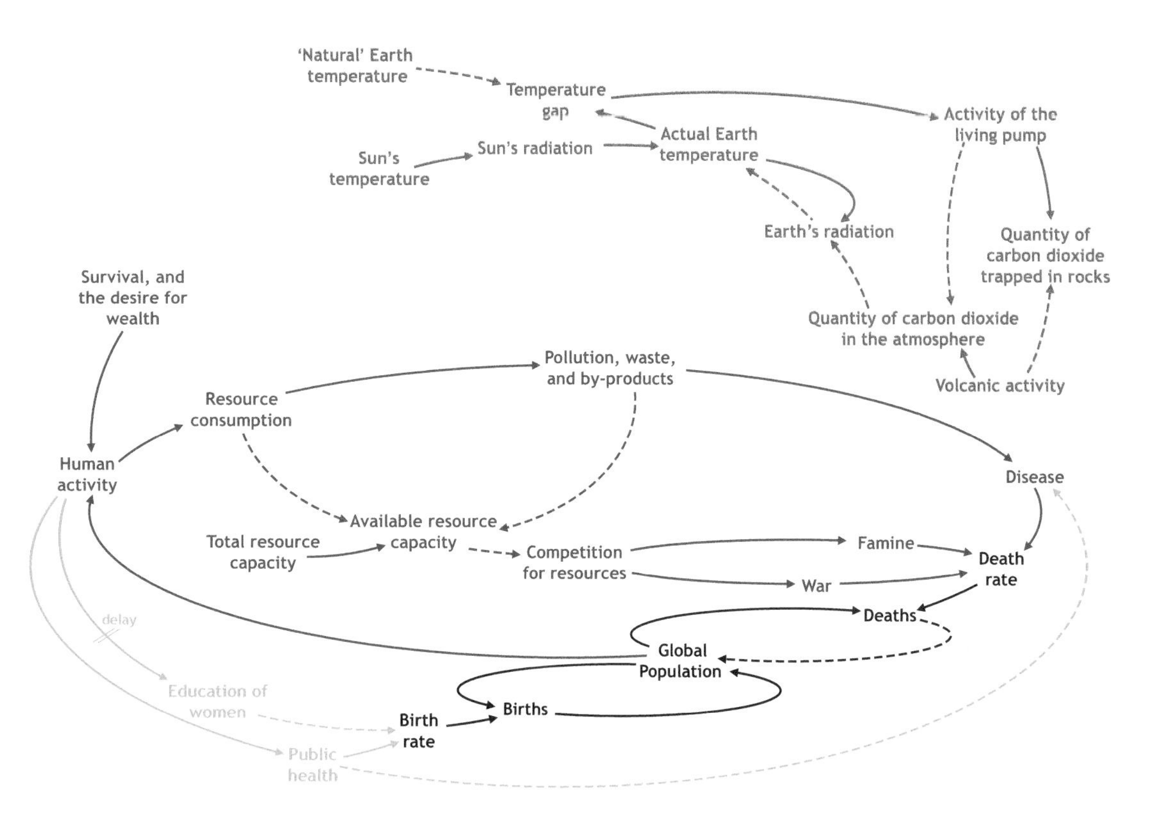

…man and Gaia lived in harmony, as two non-interacting systems…

…but Then…

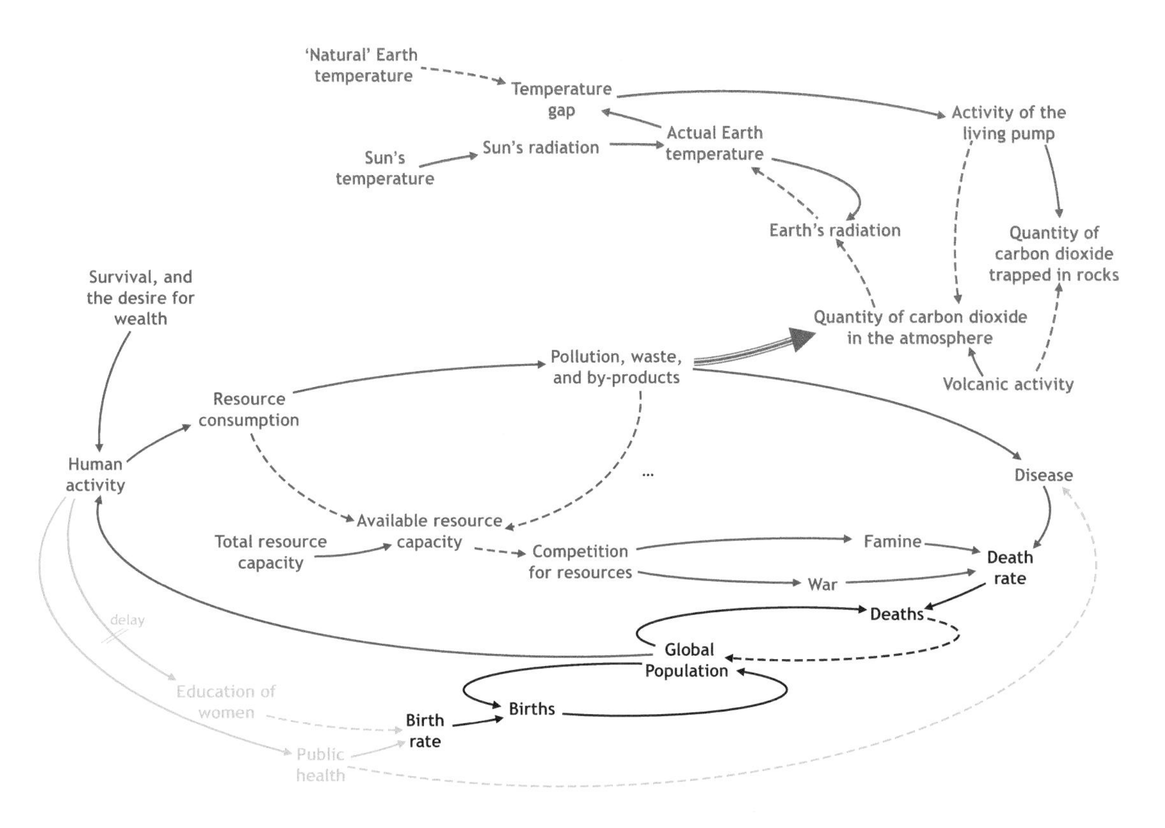

…until man-made pollution, especially carbon dioxide emissions from burning fossil fuels, increased the quantity of carbon dioxide in the atmosphere so much that the activity of the living pump can no longer cope.

The living pump just can't extract carbon dioxide fast enough, so the actual Earth temperature must inevitably rise.

Global Warming Is Real...

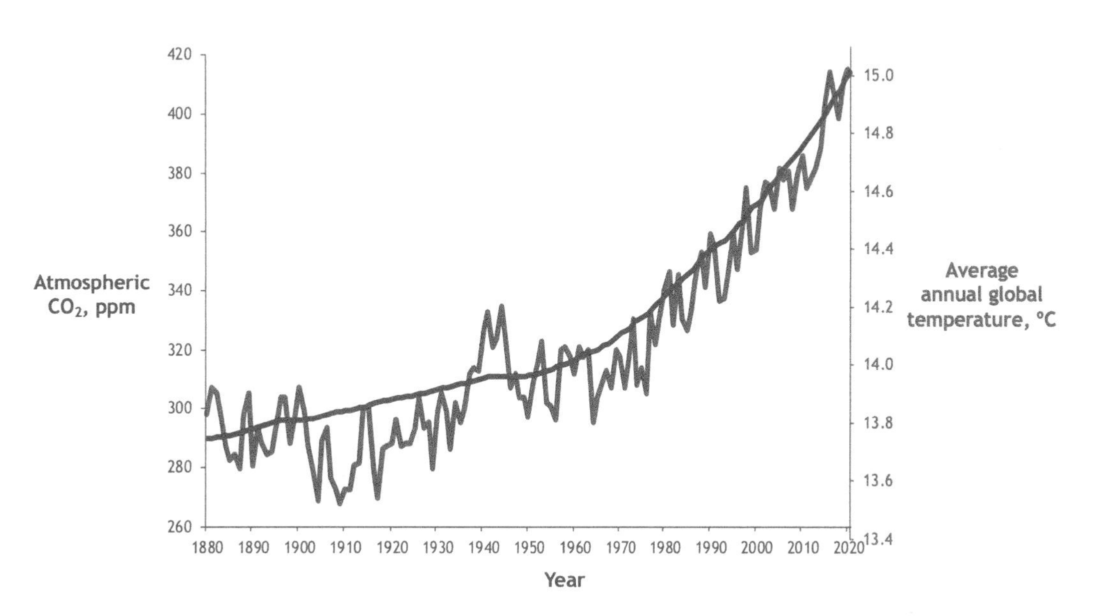

Sources: CO2 - US National Oceanic and Atmospheric Administration; Temperature - NASA Goddard Institute for Space Studies

…Causing Gaia to React…

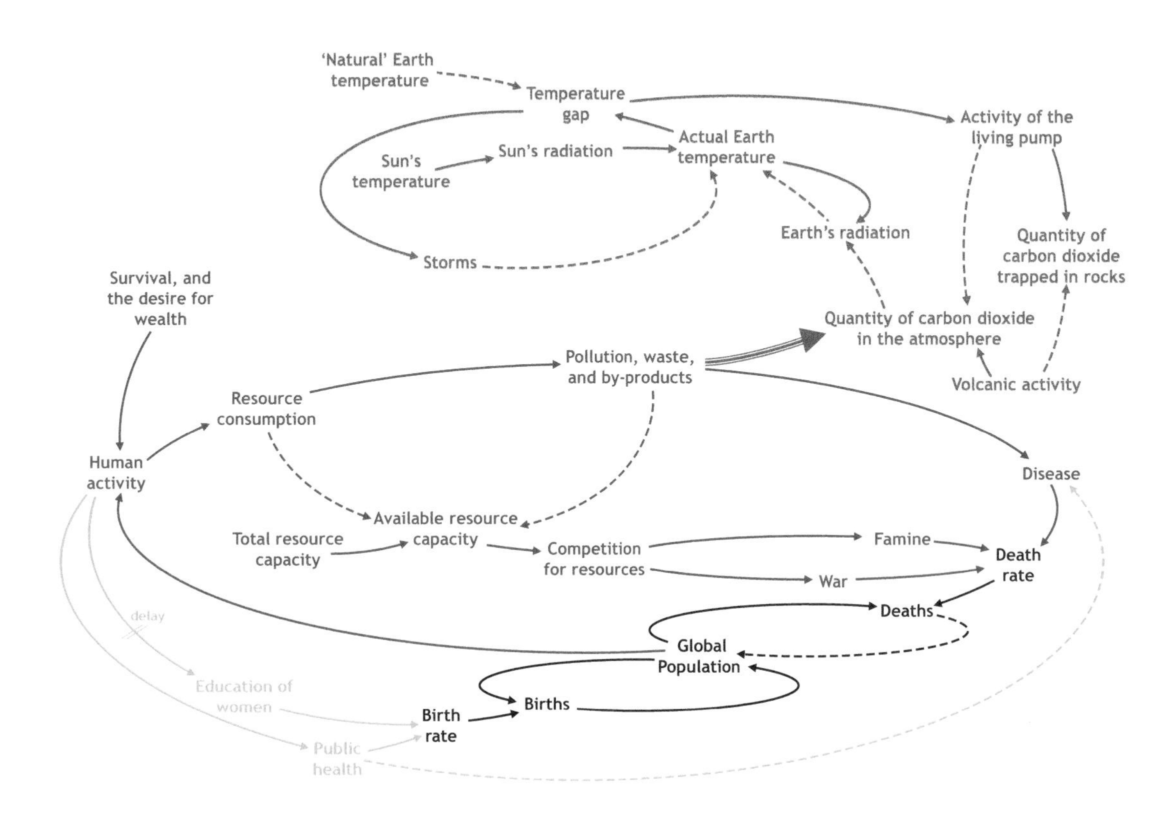

For narrative, see following page

...Causing Gaia to React...

Many living systems have more than one way of maintaining stability. In our bodies, for example, if we become hot, we first sweat, and if that doesn't lower our temperature sufficiently, our bodies then enhance the flow of blood to our skin to increase heat loss.

Gaia, too, has more than one way of maintaining the Earth's temperature stable. The *activity of the living pump* is the principal way, but if the pump just can't pump fast enough, and the *actual Earth temperature* begins to rise, then another mechanism is triggered: an increase in the incidence of *storms* and violent weather, for these act to dissipate energy, so reducing the *actual Earth temperature.*

Even though many living systems have more than one way of maintaining stability, ultimately, they break. As we have seen, if we get hot, we sweat, and increase the supply of blood to the skin so as to maximise heat loss. But if we get too hot – heatstroke, for example – we die. Gaia too has several mechanisms that can maintain the Earth's temperature stable – but they too have limits, beyond which Gaia, and the entire planet, will die.

…and to Fight Back…

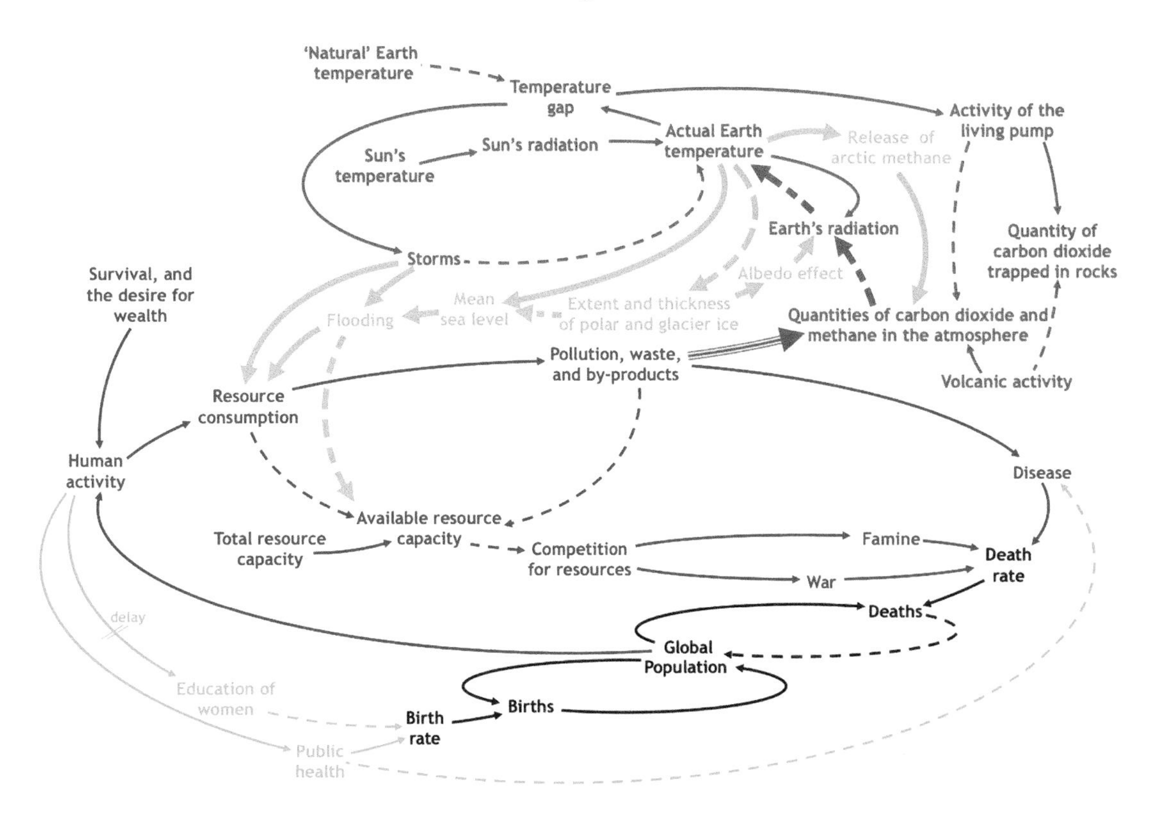

For narrative, see following page

...and to Fight Back...

An increase in the *actual Earth temperature* has a significant impact on regions of the world, which, over the last many millennia, have been very cold.

Arctic permafrost, lakes and seas contain trapped *methane*, which is progressively *released* as the *actual Earth temperature* increases. Although methane has a lower atmospheric concentration than carbon dioxide, and remains in the atmosphere for a shorter time, it is substantially more powerful as a greenhouse gas – so a sudden release of large quantities of methane, resulting from the reinforcing loop from *actual Earth temperature*, through *release of permafrost methane, quantities of carbon dioxide and methane in the atmosphere* and *Earth's radiation* back to *actual Earth temperature,* could be truly catastrophic.

Also, as the *actual Earth temperature* increases, polar ice sheets and mountain glaciers progressively melt, and the *extent and thickness of polar and glacier ice* reduces.

This has two consequences. The first concerns the *albedo effect,* the phenomenon whereby light-coloured surfaces reflect the sun's energy, in contrast to dark-coloured surfaces, that absorb energy. Ice sheets are white, and so the greater their extent, the greater the *earth's radiation*; conversely, the greater the extent of darker land and sea, the lesser the *earth's radiation*, with the absorbed energy acting to increase the *earth's actual temperature.* The melting of ice therefore ultimately replaces a white, reflecting, surface by a darker one, which absorbs more heat, causing the *Earth's actual temperature* to rise, so *melting even more ice...* as shown by the reinforcing loop from *actual Earth temperature* through *extent and thickness of polar and glacier ice, albedo effect* and *Earth's radiation*, back to *actual Earth temperature.*

Second, the melt waters from land-based ice cause a rise in *mean sea level* – something that also happens directly from the rise in the *actual Earth temperature* resulting from the thermal expansion of the water in the surface levels of the oceans. Note that the melting of the North Polar ice cap won't affect sea levels – the ice around the North Pole is floating on the waters of the Arctic Ocean, and as it melts, the resulting water 'fills' the volume previously occupied by ice.

The rise in *mean sea level* results in the *flooding* of lower-lying areas, such as Britain's East Anglia, the Netherlands and Northern Germany, Bangladesh and many islands. Flooding is also aggravated by the effects of *storms. Flooding* directly reduces the capacity of key resources, such as land and drinking water; *flooding* and *storms* both cause an increase in *resource consumption* as increasingly scarce resources are used to build, for example, sea defences to stop the flooding, and to re-build houses after a severe storm or flood.

Man's disruption of Gaia, caused by the *pollution*-driven increase in the *quantities of carbon dioxide and methane in the atmosphere*, as indicated by the thick brown arrow, has the effect of causing Gaia to fight back, as indicated by the heavy blue arrows from *storms, mean sea level* and *flooding* to *resource consumption* and *available resource capacity.*

...with a Vengeance

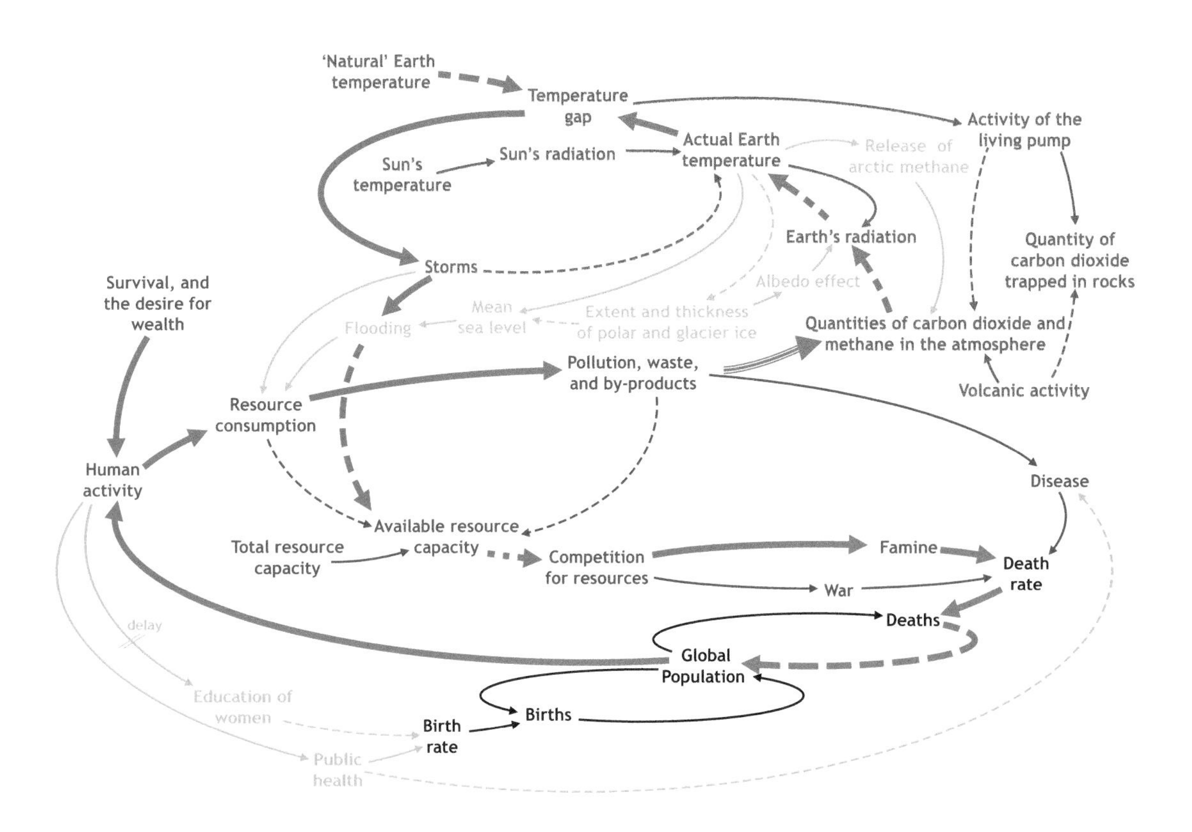

For narrative, see following page

...with a Vengeance

In the diagram on page 282, follow the closed figure-of-eight-shaped loop highlighted in magenta. There are five inverse links – from *flooding* to *available resource capacity*, *available resource capacity* to *competition for resources*, *deaths* to *population*, *quantities of carbon dioxide and methane in the atmosphere* to *Earth's radiation*, and *Earth's radiation* to *actual Earth temperature*. This loop is therefore a balancing loop – a balancing loop that seeks to converge on a target.

But what is the target?

Targets are defined by target dangles, of which there are two – the *'natural' Earth temperature*, and man's *desire for wealth*. These two targets are in competition: if man's *desire for wealth* dominates, then the *actual Earth temperature* must rise; but if the *'natural' Earth temperature* dominates, then the result, inevitably, is an increase in *deaths*.

Yes. The action of this loop is to increase *deaths* by virtue of the *famine* resulting from the depletion of the global *available resource capacity* caused by the devastation of land and fresh water supplies resulting from *flooding*. The *flooding* is caused by *storms*, in turn caused by an increase in the *temperature gap* attributable to the rise in the *actual Earth temperature* driven by an increase in the *quantities of carbon dioxide and methane in the atmosphere*, caused by *pollution*, caused by... *man* (meaning, of course, the human species). So, by increasing *deaths*, and therefore causing a decrease in *population*, Gaia is directly addressing the cause of its problem...

Just as man will swat an annoying insect, Gaia will deal with whatever it finds annoying. Gaia, after all, has existed for far, far longer than man, and if man were no longer to exist, life in other forms will continue. Gaia does not need man. But man surely needs Gaia.

The loop highlighted in magenta is not the only loop acting in this way. For example, there is another through *disease* (the threat, for example, to clean water); another through *war* (if a nation is threatened by floods, what are all those people going to do?); and another through *mean sea level*.

What do we conclude? One very sobering thought.

If man seeks to 'take on' Gaia, man sill surely lose.

Gaia will solve its own problem – by depleting the human population – of its own accord.

Unless we act first.

Some Good Things to Do…

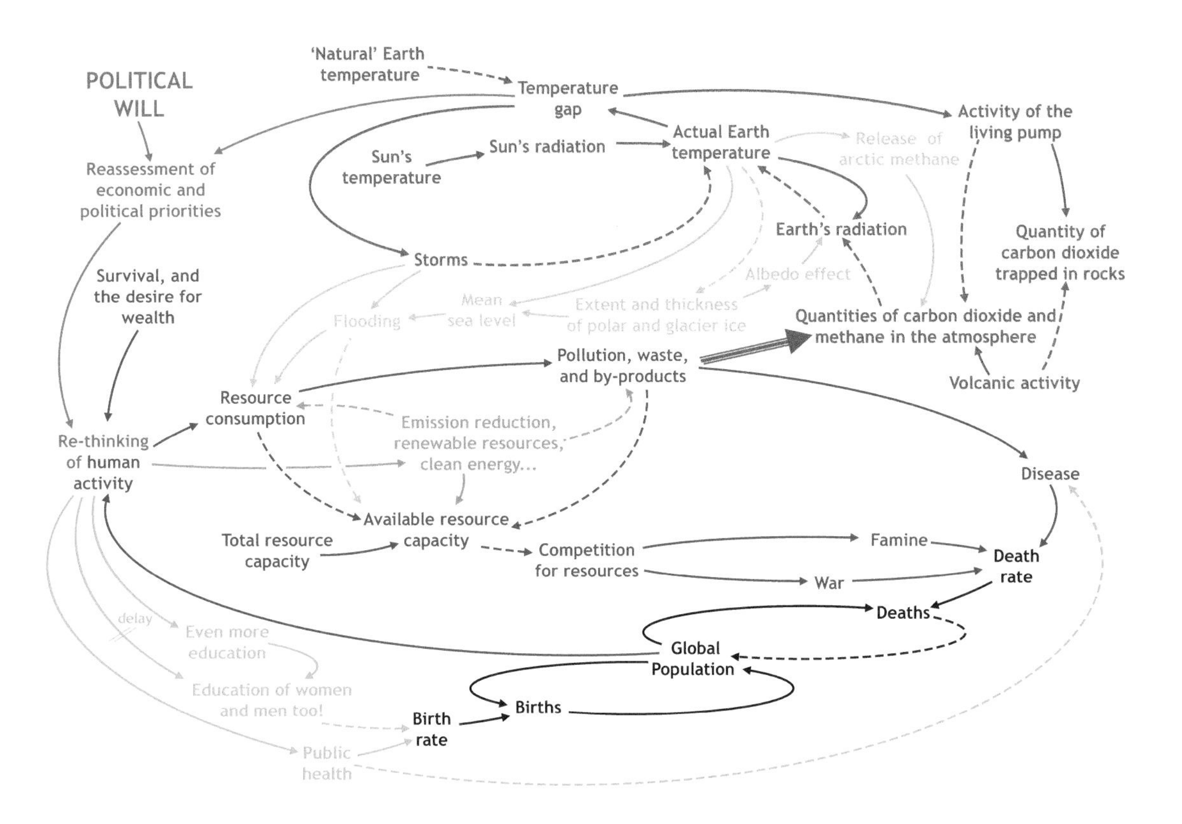

For narrative, see following page

...but Not the Right Things to Do...

To survive, man must find ways to counteract the destructive feedback loops shown on page 282. Fundamental to this is the need for the *political will* to recognise that the observed *temperature gap* is a trigger for a *radical reassessment of economic and political priorities* so as to drive a fundamental *re-thinking of human activity*...

...first, and very importantly, by enhancing the *education of women – and men too*, this being by far the most powerful method of controlling the *birth rate*...

...and second, by implementing policies for *emissions reduction, renewable resources, clean energy* and the like, so reducing overall *resource consumption*, increasing the *available resource capacity*, reducing the *competition for resources* and reducing *pollution, waste and by-products*, especially the emissions of greenhouse gases in general, and carbon dioxide in particular.

Indeed, targets for reducing emissions have been a central feature of all the major national and international initiatives, such as the 1997 Kyoto Protocol, the 2009 Copenhagen Accord and the 2016 Paris Agreement.

Reference to the causal loop diagram shown on page 284, however, will show that although reducing emissions is undoubtedly a good thing to do, it is not the right thing to do.

Why so? Because the *actual Earth temperature* is determined by the actual *quantities of carbon dioxide and methane in the atmosphere* at any time. Technically, these *quantities* are known as 'stocks' in that they accumulate over time, rather like the quantity of water in a bath.

Emissions, by contrast, are a 'unidirectional inflow' (see pages 56 and 312), causing the corresponding stock to accumulate – just like the flow of water through a tap, filling the bath. Reducing emissions *just makes the tap run more slowly* – but the *actual quantities of carbon dioxide and methane in the atmosphere* still continue to increase, as does the *actual Earth temperature*, albeit more slowly.

Reducing emissions, though a good thing to do, does not solve the climate crisis problem.

So what does?

The answer is 'geoengineering'.

Geoengineering

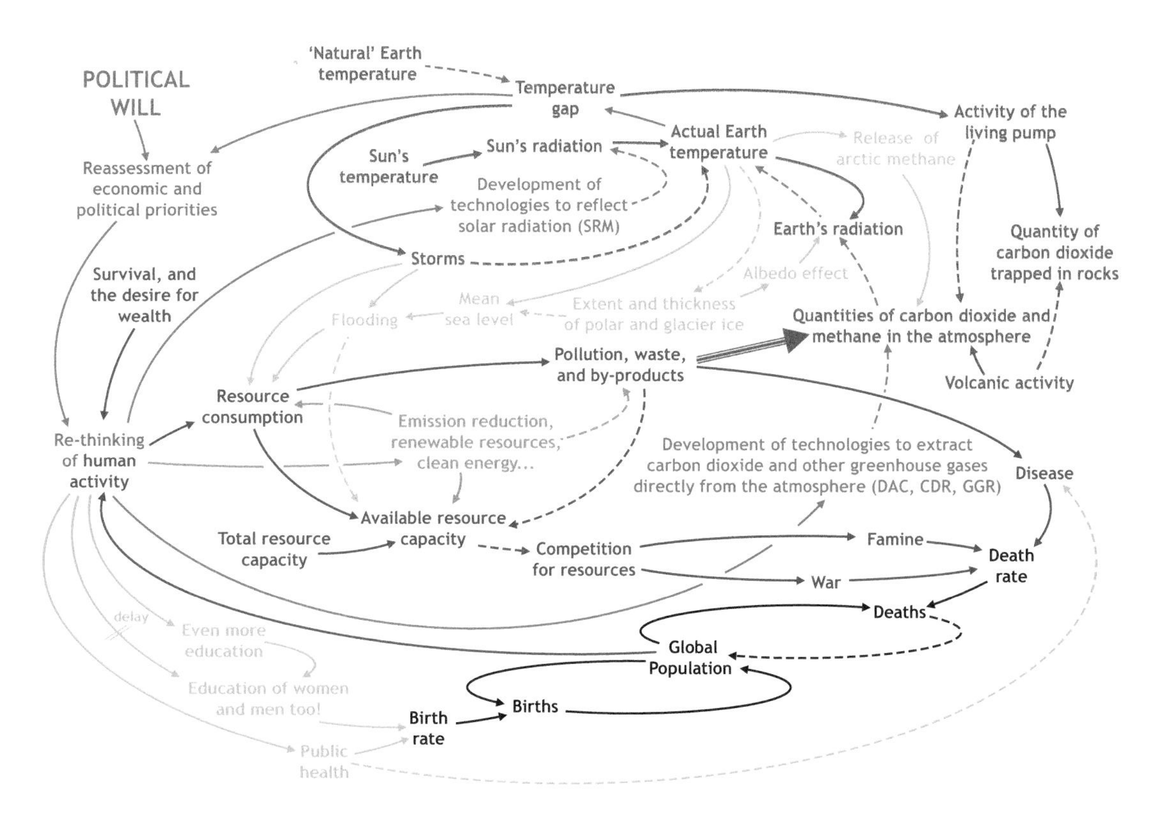

For narrative, see following page

Geoengineering

Geoengineering is the overall term for large-scale technologies that directly affect the atmosphere and the climate – technologies such as *solar radiation management (SRM), direct air capture (DAC), carbon dioxide removal (CDR)* and *greenhouse gas removal (GGR).*

The objective of *SRM* is to reduce the quantity of solar radiation that strikes the Earth's surface over any time – for example, by applying suitable coatings to the sun-facing surfaces of clouds so as to make them more *reflective* of sunlight.

The objective of *DAC*, which includes both *CDR* and also *GGR*, is to *extract carbon dioxide (and greenhouse gases in general) directly from the atmosphere.*

In principle, *CDR* mimics the natural action of coccolithophores (see page 270 and 271), which extract carbon dioxide from the atmosphere, transforming it into inert limestone or chalk – a much more permanent sequestration of atmospheric carbon dioxide than that achieved by plant photosynthesis, which forms sugars and other organic products.

CDR therefore supplements Gaia's 'living pump', and, if carried out at sufficient scale, could in principle *reduce* the *actual quantities of carbon dioxide and methane in the atmosphere* and so cause the *actual Earth temperature* to decrease.

Overall, *DAC, CDR* and *GGR* are the only approaches that will not only solve the climate crisis, but could enable the *actual Earth temperature* to be controlled. And as long as the rate of 'pumping' achieved by *DAC, CDR, GGR* and the living pump collectively is just greater than the rate of emissions, then it does not matter what the rate of emissions actually is…

The Policy to Save the Planet

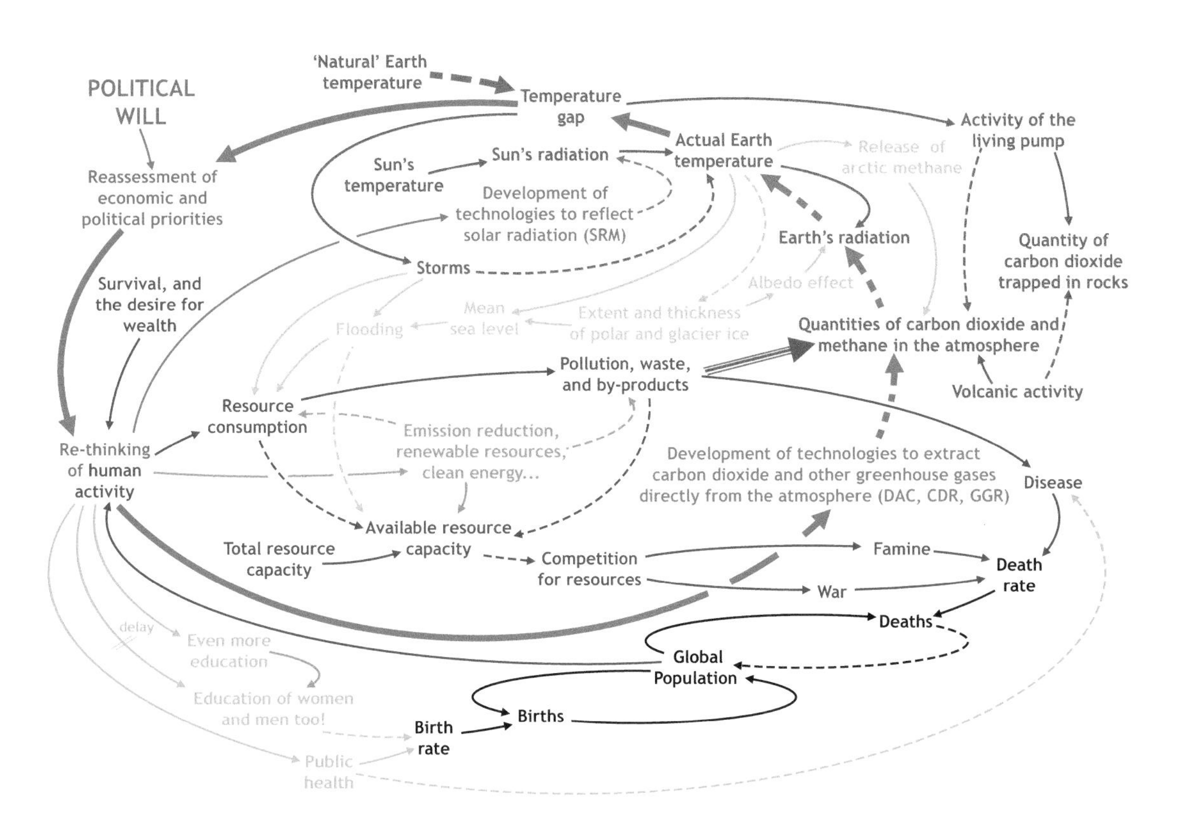

For narrative, see following page

The Policy to Save the Planet

Fundamentally, *there is too much carbon dioxide in the atmosphere now* – that's why the living pump can't cope. So cutting back on emissions isn't enough.*

Certainly, reducing emissions will help stop the problem from getting worse, but it can't solve the problem that's already there. It's rather like a ship that has sprung a leak, and has a considerable quantity of water already in the hold. Staunching the flow of water into the hold is a sensible thing to do, but if there is already too much water on board, the ship will sink, even if the hole is plugged. As well as ordering that the inflow be staunched, the wise captain *also* orders 'all hands to the pumps', for the captain knows that the water already in the hold must be baled out. And the captain also knows that, as long as the crew can pump the water out faster than the leak is letting it in, the ship is safe.

The analogy is apt, for it's all about pumps. And in the Earth's case, the 'water in the hold' is the total *quantity of carbon dioxide and methane already in the atmosphere*, the 'leak' maps on to greenhouse gas emissions, and the 'pump' is the living pump, enhanced by technology so that it has a much greater capacity. This is shown by the balancing loop highlighted in magenta on page 288, which stabilises the *actual Earth temperature* on Gaia's *'natural' Earth temperature*, 14°C.

We must develop DAC, CDR and GGR.

And we must do this now.

* More accurately, reducing emissions is an effective policy only if the rate of emissions produced (currently some 9×10^9 tonnes of carbon injected into the atmosphere per year) is consistently less than the maximum rate of the living pump (estimated at about $1–2\times10^9$ tonnes of carbon removed from the atmosphere per year). This, however, requires emissions to be reduced by at least 75% (and possibly as much as 90%): a reduction far beyond – as vividly demonstrated at the Copenhagen 2009 Climate Change Conference – any feasible economic or political possibility.

'Carbon Dioxide Removal Techniques Address the Root Cause of Climate Change by Removing Greenhouse Gases from the Atmosphere'

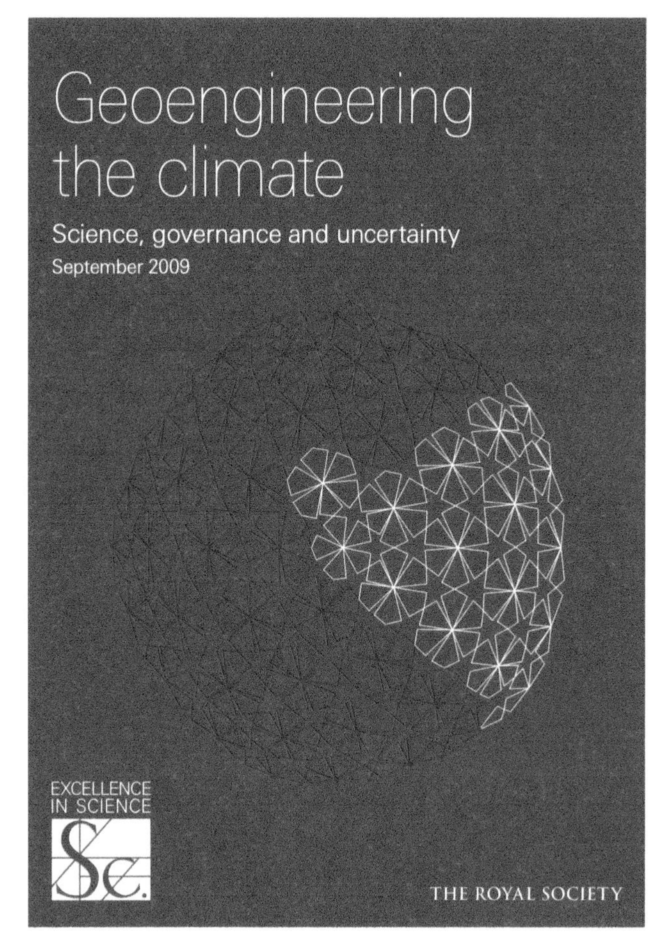

Source: https://royalsociety.org/~/media/royal_society_content/policy/publications/2009/8693.pdf, page 4

PART 3 Over to You!

Chapter 18

How to Draw Causal Loop Diagrams

DOI: 10.4324/9781003304050-21

How to Draw Causal Loop Diagrams

So far, this book has presented a wide range of examples of how systems thinking in general, and causal loop diagrams in particular, can tame complexity. The systems described, however, have been 'someone else's'. Which, I trust, has been interesting and informative, and has whetted your appetite to start drawing your own causal loop diagrams, diagrams that tame the complexity of systems that are relevant, and important, to you. So the purpose of this chapter is to offer some guidance – and indeed encouragement – to do just that.

Some of diagrams presented might be directly relevant, or might be easily modified to make them so; there might also be some 'building blocks' that can be used within a more complex picture – for example, the 'MINIMUM' structure, that describes *satisfied demand, unsatisfied demand* and *surplus capacity,* as discussed on pages 101, 109, 110 and 114, has very broad applicability.

That's helpful, but please don't expect the examples in this book to be fully applicable to your own circumstances. This chapter therefore tables my 12 rules for drawing good causal loop diagrams, rules that I use myself. The rules are shown on the next page, and explained subsequently...

...but before you can apply the rules, you need to have the right equipment.

That's not too difficult. And I'm about to show my age... for the three most important items are:

- Lots, lots and lots, of paper...
- Some sharp pencils, and
- The biggest waste-paper basket you can find.

The waste-paper basket is to catch all the (huge number of) pages that will be discarded with scratchings, doodles, curly arrows, and bits of things that really don't work. For drawing a good, insightful causal loop diagram, is A LOT OF WORK, work which isn't 'right' until right at the end. And on the journey to the 'end', everything along the way is not 'right', but decidedly wrong. Hence the waste paper basket. Yes, the causal loop diagrams presented in this book are all neat and tidy, and – I hope! – quite insightful too. But the amount of stuff I threw away...

Twelve Rules for Drawing Causal Loop Diagrams

Rule 1: Know your boundaries

Rule 2: Start somewhere interesting

Rule 3: Ask 'What does this drive?' and 'What is this driven by?'

Rule 4: Don't get cluttered

Rule 5: Use nouns, not verbs

Rule 6: Don't use terms such as 'increase in' or 'decrease in'

Rule 7: Don't be afraid of 'unusual' items

Rule 8: Designate, correctly, all links as you go along

Rule 9: Keep going

Rule 10: A good diagram must be recognised as real

Rule 11: Don't fall in love with your diagrams

Rule 12: No diagram is ever 'finished'

Boundaries

Rule 1: Know your boundaries

It's helpful, at the outset, to take a view on the scope of interest, of what's 'inside' and what 'outside'. So, for example, is the system about a particular activity (say, budgeting), or the activities of the organisation as a whole? Or of the organisation in the context of suppliers, customers and competitors?

A 'clue' to a system's boundaries are the dangles, for, by definition, in any one causal loop diagram, any input dangle acts as a 'driver' of the system, and any output dangle is the outcome of the system's operation. So the input dangle *target sales revenue* will drive a balancing loop in which the difference between the *target sales revenue* and the *actual sales revenue* – the *sales revenue gap* – will drive whatever actions are appropriate.

Two related points. First, for a particular individual, any *target* or *goal* is often set by someone more senior, and therefore regarded as a 'given', determined by someone outside, beyond 'my' system boundary, and therefore an input dangle; for the more senior manager, however, this item is likely to be one of several output dangles, outcomes of a system driven by an input dangle such as *company target profit* or perhaps *required returns to shareholders.*

And second, as already noted, output dangles represent the outcomes of the system's operation – what actually happens. This is not – as some might believe – a manifestation of the system's *purpose*: as discussed on page 10, systems don't have a 'purpose'; rather, they deliver outcomes. A system's purpose is not a property of the system, but rather the intent of the person who designed the system. And the extent to which the actual system outcome matches the designer's purpose is an indication of the quality – or otherwise – of that system's design.

In general, any one causal loop diagram will contain many variables, but relatively few input and output dangles. Collectively, these dangles define the boundary of the system, and the values of the input dangles determine the values of the output dangles – all the other, many, internal variables 'look after themselves'.

Where to Start

Rule 2: Start somewhere interesting

In any casual loop diagram, ultimately, everything is connected to everything else – so in principle, it doesn't matter where you start, for sooner or later, everything becomes 'joined up'.

In practice, however, starting in an 'interesting' place is often very helpful.

What might be 'interesting', of course, is dependent on the context, but here are two suggestions.

First, whenever a systems thinking study is initiated, there is usually something in mind, something that's important to understand more richly. That 'something' (or those 'somethings') will often identify a small number of key variables that could be very good places to start, for they have already been identified as significant.

Second, it is often very helpful to identify, early on, any relevant *targets, goals, objectives* or *ambitions* and use those as starting points, confident in the knowledge that each is driving a balancing loop.

What's Next? What's Before?

Rule 3: Ask 'What does this drive?' and 'What is this driven by?'

In that everything is connected to everything else, given any variable, these two questions take you forwards, from cause to effect, and backwards, from an effect to a preceding cause.

In practice, there might be multiple answers to both questions, which is fine – note them all down. So, for example, the variable *my workload* might be considered to drive, for example, *stress, likelihood of error* and *possibility of resignation,* in that all these (and, quite likely, some other things too) are plausible consequences of an increase in *my workload.*

The next task is to organise these as plausible chains of causality: are the three results independent, suggesting a structure such as that shown below on the left, or is *stress* the immediate consequence of an increase in *my workload,* with the *likelihood of error* and *possibility of resignation* driven by *stress,* as on the right?

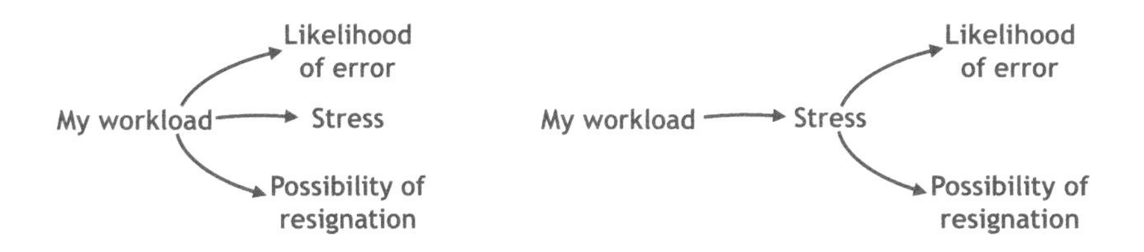

Or perhaps something else.

So that needs thinking through, always going back to the fundamental story that the causal loop diagram describes, for all diagrams must tell a true story.

And if you're not sure, or if you've missed something out or included something that shouldn't be there – *this does not matter right now.* You will spot things, and change things, later. That's what the waste paper basket is for.

Beware Superfluous Detail

Rule 4: Don't get cluttered

Good causal loop diagrams are sparse, but insightful, with fewer, rather than more, variables – and the 'right' ones, as required to tell the story.

But in compiling a diagram, it is very easy to include unnecessary – and distracting – clutter.

And there are three reasons why this happens.

First, the feeling that 'this might be important, I'd better leave it in – after all I can always take it out later'. Yes, it can be taken out later, and much is taken out later – one of the last tasks when reviewing a 'good version' is to ask 'is this variable really necessary?'. It can also be put in later, too – if it is found to be necessary, so if there is some doubt, *leave it out* – but keep a note of it, so that it's not forgotten, and then, later, review the list of 'items thought about but not yet included' to determine whether it would be better to add it back in.

Second, we are all used to detail, especially if we use spreadsheets in general, and accounting spreadsheets in particular. When doing the accounts, the unit price of [this component] matters. But in a causal loop diagram which includes the variable the *total costs of production,* the *price* at which a particular component is purchased is probably a detail too far.

Third, gathering all that detail, writing it down and drawing all those curly arrows are 'things to do', keeping you busy, and giving the feeling that 'yes, I'm really making progress!'. Yes, doing all that is indeed something to do. But in the context of drawing an insightful causal loop diagram, it might not be making progress at all…

Nouns, Not Verbs

Rule 5: Use nouns, not verbs

A feature of every causal loop diagram in this book is that all the variables are all either nouns – such as *sales* – or noun phrases – such as *development of technologies to extract carbon dioxide and other greenhouse gases directly from the atmosphere (DAC, CDR, GGR).*

Importantly, there are no stand-alone verbs (*sell*) or verb phrases (*develop learning culture*).

The use of nouns rather than verbs has the benefit of throwing the emphasis on outcomes, on things that actually happen.

Certainly, to make [whatever it is] happen requires an action, which is most naturally described using a verb. The action, however, is transient; the result of that action more permanent. So using nouns rather than verbs creates a more tangible 'mental picture' of what the world 'looks like' now, or might 'look like' in the future, rather than on the process of getting there.

Accordingly, I suggest that it is advisable to use nouns, not verbs, for all variables. And as I trust this book has demonstrated, this is easy to do, and results in clear diagrams.

That said, there is one frequent occurrence in which a verb appears to be more 'natural' to use than a noun – and that's in a balancing loop associated with a target dangle. As discussed on pages 67 to 70, in order to close the *gap* between a given *target* and the corresponding *actual*, some appropriate *action* needs to be taken, and actions are of course most readily described by verbs. So, for example, the *headcount gap* between my *target headcount* and my *current actual headcount* leads to actions associated with verbs such as *hire* and *fire*. Indeed; but the nouns *hiring and firing* work just as well, so the 'use nouns only' principle works here too.

'Increase' and 'Decrease' Are Inherent in the Links

Rule 6: Don't use terms such as 'increase in' or 'decrease in'

Yes, an increase in the *number of staff* does drive an increase in the *organisational capacity for work*. And this truth makes it very natural to compile a causal loop diagram showing a variable named *increase in the number of staff* connected to a variable named *increase in organisational capacity for work* by a direct link.

In general, such a temptation is best avoided.

In this example, the variables of interest are the *number of staff* and the *organisational capacity for work*, and the use of a direct link necessarily implies that an increase in the one will drive an increase in the other. So to include '*increase in*' with the variable description is unnecessary. Furthermore, to show the variables as *number of staff* and the *organisational capacity for work* connected by a direct link also allows for the possibility that a decrease in the *number of staff* will drive a decrease in the *organisational capacity for work* – which is somewhat obscured if the variables are described as *increase in...*

More subtly, a variable described as *number of staff* is in fact different from a variable described as *increase in number of staff* – and especially so when that is intended to mean *increase in number of staff per year*, but with the *per year* left out and assumed to be understood. So, for example, the *number of staff* is an instantaneous measure at any specific time – say, 25, now. In contrast, the *increase in number of staff per year* is a measurement over a period of time, and determines the rate at which a particular item changes. A variable (say, the *number of staff*) and that variable's rate of change (the *increase in number of staff per year*) are different, but it is very easy to get them confused.

Sometimes, a causal loop diagram needs to include a 'rate of change' variable in its own right, perhaps in addition to the 'underlying' variable too. If this is required, then it is wise to distinguish clearly between them, as, for example, two variables named *number of staff* and *rate of change in number of staff, per year* – so avoiding 'increase in' and 'decrease in', whilst being totally clear as to what, precisely, each variable is.

There are, however, some instances in which incorporating directionality within a variable name is acceptable: for example, the *pressure to decrease capacity* appears in the causal loop diagram on page 111 in relation to hospital admissions. This *pressure* may be strong or weak, and necessarily acts to *decrease capacity*, for it is driven by a *surplus capacity*, which is either zero or a positive number. Likewise, the *pressure to increase capacity*, driven by the number *of patients not treated*, in the same diagram, and also the *downwards pressure in price* and *upwards pressure on price*, driven by *excess supply* and *unsatisfied demand*, respectively, in the causal loop describing *supply, demand* and *price* on page 118.

'Unusual' Items Are Usually Very Real

Rule 7: Don't be afraid of 'unusual' items

An important feature of causal loop diagrams is that they tell real stories, about what actually happens.

So variables such as *ambition to meet performance measures*, *stress*, *pressure to do [this]* and *well-being* will be features of many diagrams, and indeed important features too, for these very human characteristics are often the principal drivers of how a real system actually behaves. And despite the significance of these items, they certainly do not feature in any organisation's formal accounts, so there is no requirement to measure them. That said, a wise organisation is alert to them, and there is increasing pressure on organisations to make statements on at least some of them in their annual reports.

There are some very important 'non-human' variables too – variables that in principle should be routinely measured, reported and acted upon. For example, the causal loop diagram on page 94 includes a 'family' of variables of the form *effect of investment in [this] on improving business performance* – these, in effect, answering the question 'By what amount will *sales revenue* increase, or *total costs* decrease, as the result of an investment of [this amount] in [whatever]?'. That question applies to every commercial enterprise which makes sales and generates profits; equivalent statements can be made for non-commercial activities such as a school or the police service.

Questions of this type are often tough, really tough. And, as a result, they are often avoided.

But as the causal loop diagram on page 94 makes vividly clear, these variables are central to the fundamental reinforcing loop of business growth. So understanding them as fully as possible is probably a sensible thing to do, for this will inform how best to answer the single most important question faced by any enterprise: 'to maximise the future success of the enterprise, what is the optimal allocation of [this amount] over these [possible opportunities for investment]?'

By documenting how systems actually work, an insightful causal loop diagram will identify all the variables that are truly important, even if they are not routinely measured, or are 'unusual'.

Direct? Or Inverse?

Rule 8: Designate, correctly, all the links as you go along

When compiling a causal loop diagram, there is often a strong temptation to write as many variables as you can think of on a page, and then draw all sorts of curly arrows, all over the place, to join things up. And an even stronger temptation to think 'I can always determine which links are direct, and which are inverse, later – much more important to get the diagram, as a whole, done'.

Yes, it is a good idea to compile a list of relevant variables early on – but do that on a separate page, not the page you are working on for the diagram. That page is very likely to be thrown away, so you don't want to lose the list too; also, by having the variables you think you are likely to use on a separate page, you can choose (and then tick off) the particular variable you need now, so building the diagram step by step.

And as you do that, it is wise to identify each individual link as direct or inverse as each pair of variables are connected. Yes, sometimes this is hard to think about. As *demand* increases, does the *delivery time* go up or down? Mmm.... quite often, the *delivery time* doesn't actually change... and when the *demand* goes down a bit, the *delivery time* doesn't usually change, we just despatch things as normal... so maybe there is no link at all... But when the *demand goes* up, substantially, we're in real trouble, and the *delivery time* goes through the roof... Ah! That's it. A direct link from *demand* to *delivery time*, and an inverse link from *capacity* to *delivery time...*

The requirement here is for very clear thinking, and that can sometimes be difficult. But systems thinking, and the compilation of succinct and insightful causal loop diagrams, can be a valuable stimulus to think clearly.

Doing the links as you go along has another benefit too. Suppose you're thinking about the link between *overtime hours worked* and *likelihood of error*. Is that a direct link (someone working overtime is quite likely to be tired, and so more likely to make mistakes) or inverse (by virtue of the additional time available to complete a task carefully)?

Both explanations are plausible, but an important rule is that all links must be either direct or inverse (with the rare exception of influence links which should only be used for dangles). So the question of which link should be used triggers thinking such as 'maybe there are some other variables that are relevant, but were not on the original list – variables such as *stress caused by time pressure'*, which might be a prime cause of errors, whether overtime is worked or not. This changes the structure of the diagram, for the better, and is the result of the discipline of designating, correctly, all the links as the diagram is being compiled, and not 'leaving it for later'.

Don't Give Up!

Rule 9: Keep going

The waste paper basket is full.

The scraps of paper that aren't in the basket are strewn with unconnected fragments, with curly arrows all over the place.

Some things make sense, but many things don't.

And you know that, somehow, the [*consumption of timber*], as stimulated by [*government incentives to use biomass as a fuel for generating electricity*], is related to [*the number of small furniture companies going out of business*], but you can't see how everything fits together... oh dear... what a mess...

Yes, it is a mess.

And it might get messier before it gets tidier.

But it will get tidier. And when you notice that the [*price of timber*] is driven upwards by the [*demand for timber as a biomass fuel*], which then increases [*the cost of raw materials to furniture manufacturers*], perhaps things start coming together...

So don't give up.

Keep going.

Even when it's tough.

Causal Loop Diagrams and Reality

Rule 10: A good diagram must be recognised as real

Sometimes, a causal loop diagram is intentionally drawn as a hypothetical future ideal – for example, to contribute to a discussion about 'suppose that the [whatever] system were to look like [this]?'

But before that happens, it is usually helpful to compile a diagram describing how the [whatever] system works now, so that everyone can agree that, yes, that's how it is. Although compiling and agreeing a good diagram of the 'as is' state might appear, to those impatient to do something new, to be wasting time, that is never the case: it is almost always true that different people who interact with the system have different mental models of how that system operates. So compiling a good causal loop diagram of how the system works now is a key step in ensuring that everyone is, quite literally, 'on the same page'.

And then, if there is agreement that the system's outcomes might be better, the diagram provides the right platform for asking, and then answering, the most important question in improving a system's operation: 'how might [this feature] be different?'. For if you want a different, and better, outcome, then something about the existing system has to be different, and so this question triggers a well-focused enquiry, which is highly likely to generate ideas.

But before those ideas are generated, the starting point is an agreed description of the world-as-it-is-today, and a good causal loop diagram is the best place to start. So it's very important that, at the outset, everyone agrees that, yes, that is indeed how our 'world' works.

A word of caution. Sometimes, someone will say, 'Well, of course things work like that! That's obvious! Why did we waste so much effort coming up with stuff we already know?'

That is an intentional put-down. But is in fact good news, for it is an acknowledgement, usually from an antagonist, that the diagram is indeed a good representation of the current system. So whenever that happens, I reply, 'Thank you. It's really helpful that you recognise the diagram as real, and as a good description of what is happening now. And perhaps it contains something that some of us haven't quite noticed before…'.

Beware of 'Love'...

Rule 11: Don't fall in love with your diagrams

However 'beautiful' they are.

Yes, it takes a long time to compile a really good diagram, and when it's finished, it's 'beauty' can sometimes be a huge barrier to changing it....

...which can lead its author to find all sorts of reasons why [*that variable*] is, yes, agreed, 'interesting', but surely not so important, and so shouldn't be incorporated – whilst thinking, deep down 'but it is important, and should be included, but I just can't face doing that diagram the all over again...'

But it is likely that the diagram will indeed have to be re-drawn, for just squeezing the new variable in, and having the links criss-crossing in an unintelligible tangle, just isn't good...

...so re-draw it. That's the right thing to do. And don't let 'love' – or laziness – get in the way.

...for Even the 'Best' Diagrams Might Become Even Better...

Rule 12: No diagram is ever 'finished'

Or indeed 'right'.

At any time, the best that can be said of any causal loop diagram is that it is a faithful representation of the mental model of how that particular aspect of the 'world' behaves, as perceived by whoever compiled the diagram.

And the key benefit is just that. It allows one person's individual – or a team's collective – mental model to be explained, clearly, to others. Other people's mental models of ostensibly the same reality might be different, and very often, when a causal loop diagram is presented and discussed, different people will have different views. Usually, however, it is possible to converge on a shared diagram *structure* – this being the definition of the variables and how they are linked together, which can often be an objective reality – but to have different opinions as to the strength of different links.

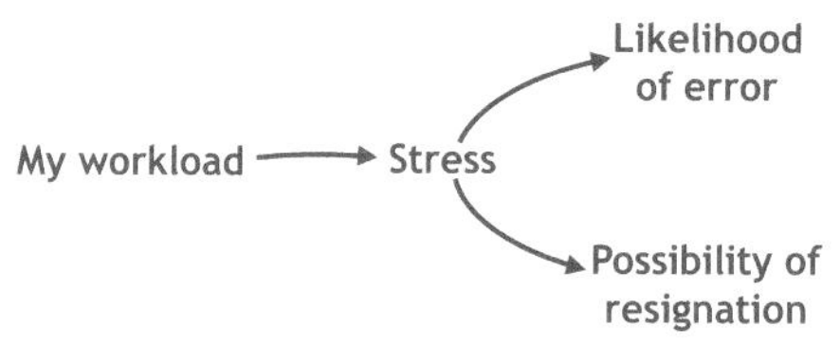

For example, we might all agree that the structure to the right is a valid description of reality, but some may believe that the link from *my workload* to *stress* is very weak, and that 'we can take it'; others may believe it to be quite strong, leading to *errors* being made by perhaps more junior staff.

That's an important organisational conversation to have, and could lead to a discussion of matters that aren't on the diagram at all, such as the process by which work is allocated to members of staff, and how the organisation does, or does not, allow 'permission' for staff to say 'I'm getting stressed' so that something can be done about it before those *errors* happen... which, in turn, might initiate a conversation about the importance, or otherwise, of an individual's work-life balance with the overall organisational culture.

This might also result in agreement to add some further features – such as *organisational 'safety' to ask for help* – into the current diagram, which, even then, won't be 'finished'...

Chapter 19

How to Use Causal Loop Diagrams

DOI: 10.4324/9781003304050-22

Stimulating Discussion

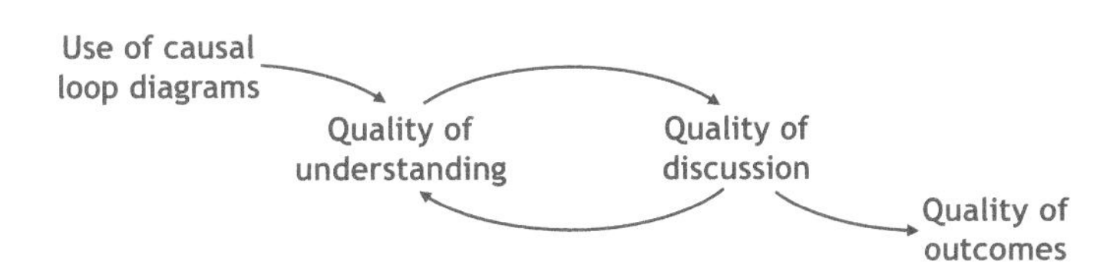

As discussed on page 9, any causal loop diagram is a representation of the originator's mental model of a given reality: 'my' diagram describes how 'I' believe the world works. As a result, 'I' will have all sorts of ideas as to what needs to happen to the corresponding system so that it works (in 'my' opinion) more effectively and delivers better outcomes.

'You' may believe that the world behaves differently, and that different interventions are required. And the differences in world-view, the differences between our mental models, can result in bitter argument – and perhaps worse.

These arguments can be acrimonious because neither participant has fully explained what their mental model actually is, and the other participant isn't listening. After all, who starts a conversation by saying, 'let me explain my mental model…'?

This is where good causal loop diagrams can really help. Suppose that both participants were to pause, and to compile their own diagrams of what they believe. That exercise alone is valuable, for whenever you compile a diagram, you inevitably identify 'loose ends' or possibilities you hadn't considered before.

And then the diagrams can be compared, not with an attitude of 'I'm right – you're wrong', but in a spirit of enquiry – 'this is what I think; what do you think?'

One of the most powerful benefits of systems thinking is in stimulating discussion, in encouraging constructive debate. Ideally, everyone can converge on the same diagram, with the same variables, linked together in the same way, for the 'world' should be the same external reality. But if different people have different views about the strengths of some of the links, then systems with the same structure will exhibit different behaviours, and result in different outcomes. Furthermore, and importantly, different people will have different views about the likely impact of different interventions… which is all about changing the structure of the system, about creativity and innovation….

Stimulating Creativity and Innovation

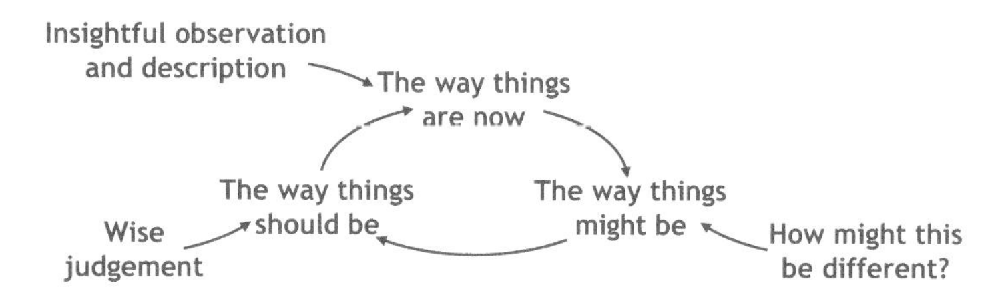

Many people believe that creativity is about the discovery of something new. No. It isn't. It's about the discovery of something *different*; *different from what is happening now, and different in a way that makes what is happening now better*. If that difference is 'new' too, that's all to the good – but it's the *difference* that matters, not any novelty.

As discussed on page 9, a good causal loop diagram, based on careful observation, is by far the most powerful way of describing how a complex system behaves, insightfully depicting *the way things are now*...

...so providing a robust platform for creativity, the discovery of ideas that open the possibility of *the way things might be,* ideas as to how the existing system might be different, and deliver better outcomes.

And to do that, the key question is *'how might this be different?'*

This question may be asked in many different ways, for example:

- Suppose [this link] were stronger, or weaker?
- Suppose [this variable] were connected differently? Or eliminated?
- What other variables might be part of the system?

Exploring these questions will generate many ideas... but not all ideas are good ones... so we need to exercise *wise judgement* to identify *the way things should be*, so that those good ideas, once successfully implemented, become *the way things are now*... and when that has happened, that virtuous circle can continue...

Stimulating Dynamic Modelling

Causal loop diagrams are necessarily static pictures of the relationships between the system's variables. That said, the structure of reinforcing and balancing loops does provide some insight into the dynamic behaviour of the system – growth, decline, convergence on a target – but this is generalised and qualitative, rather than specific and quantitative. In many circumstances, the insight obtained from a causal loop diagram is sufficient to ensure a good discussion, and to spur creativity; sometimes, however, there is further benefit to be obtained by answering questions such as 'how long is it likely to be until...?', 'what is the likely maximum value of...?' and 'what needs to done to ensure that [this variable] is highly unlikely to exceed [this value]?' – questions of great importance during the Covid-19 pandemic.

To obtain quantitative estimates, we need to use a **system dynamics computer simulation model**, which enables the behaviour of the system to be tracked over time. A key concept in simulation modelling is to distinguish between three different types of variable, known as **stocks** (representing variables that accumulate over time), **flows** (variables that increase or decrease the corresponding stocks) and **converters** (variables most appropriately represented by a current number, even though, in principle, they may be either stocks or flows) as represented in what is known as a **'stock-flow diagram'** of which this is an extremely simple example:

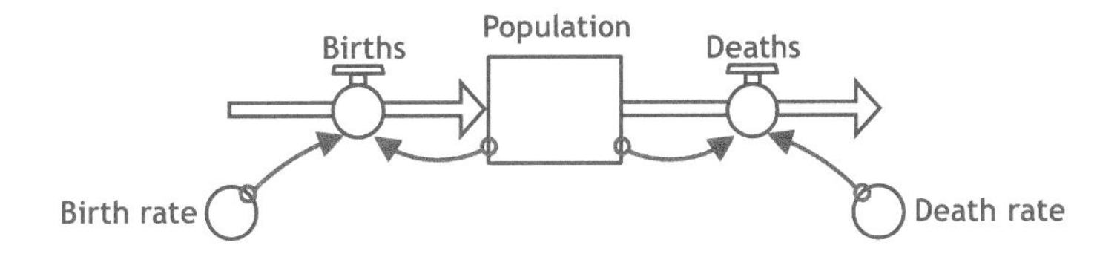

Compiling a good causal loop diagram can often serve as the specification for a system dynamics model, for the structure of the model must reflect the structure of the causal loop diagram. As an example, the system dynamics model shown in the illustration on this page corresponds to causal loop diagram on page 58, with the *population* being modelled as a stock, *births* as a unidirectional inflow (so increasing the *population* stock), *deaths* as a unidirectional outflow (so decreasing the *population* stock), and the *birth rate* and *death rate* as converters.

To give informative results, a system dynamics model must specify all the mathematical relationships that define the action of all the links that connect the variables in the corresponding causal loop diagram. This usually requires very clear thinking, as well as considerable research, often introducing many additional variables that are not required in the causal loop diagram. Specifying, writing, validating and using a system dynamics model is a demanding task, further details for which will be found in some of the books, and other resources, suggested on the following pages.

Some Good Things to Read and Some Resources Too

Some Good Things to Read…

On systems thinking

Seeing the Forest for the Trees: A Manager's Guide to Applying Systems Thinking, by Dennis Sherwood, published by Nicholas Brealey Publishing, 2002.

Thinking in Systems: A Primer, by Donella Meadows, edited by Diana Wright, published Chelsea Green Publishing Co., 2017.

The Fifth Discipline: The Art and Practice of the Learning Organisation, by Peter Senge, published by Random House Business, 2nd edition, 2006.

On systems thinking and system dynamics simulation modelling

Business Dynamics: Systems Thinking and Modelling for a Complex World, by John Sterman, published by McGraw-Hill Education, 2000.

Strategic Modelling and Business Dynamics, by John Morecroft, published by Wiley, 2nd edition, 2015.

Competitive Strategy Dynamics, by Kim Warren, published by Wiley, 2002.

On Gaia…

Gaia: The Practical Science of Planetary Medicine, by James Lovelock, published by Gaia Books Limited, 1991.

On creativity

Smart Things to Know about Innovation and Creativity, by Dennis Sherwood, published by Capstone Publishing, 2001.

Creativity for Scientists and Engineers, by Dennis Sherwood, published by the Institute of Physics, 2022.

How to Be Creative – A Practical Guide for the Mathematical Sciences, by Nicholas J Higham and Dennis Sherwood, published by SIAM, The Society for Industrial and Applied Mathematics, 2022.

On behavioural economics

Nudge: The Final Edition, by Richard Thaler and Cass Sunstein, published by Penguin, 2021.

Thinking, Fast and Slow, by Daniel Kahneman, published by Penguin, 2012.

Some Resources

For drawing causal loop diagrams

All the causal loop diagrams in this book have been drawn using PowerPoint; here are some software products that have been specifically designed for that purpose:

Loopy, designed by highly imaginative Nicky Case – for more details see https://ncase.me/loopy/.

Cauzality, an on-line mapping tool, suitable for use by teams – for more details see https://cauzality.com/.

VisualParadigm's causal loop diagram tool – for more details see https://online.visual-paradigm.com/diagrams/features/causal-loop-diagram-tool/.

Creately's causal loop diagram template – for more details see https://creately.com/diagram/example/jqq5komr4/Causal+Loop+Diagram.

Vensim, a system dynamics modelling software package, also draws causal loop diagrams – https://vensim.com.

Software for system dynamics simulation modelling

Stella, and *ithink*, both available from isee systems – https://www.iseesystems.com/.

Vensim, from Ventana Systems – https://vensim.com/.

Index

T - #0074 - 131222 - C360 - 210/280/16 - PB - 9781032302331 - Gloss Lamination